AF765599

HISTORY OF IRRIGATION, DRAINAGE AND FLOOD CONTROL IN NIGERIA FROM PRE-COLONIAL TIME TO 1999

HISTORY OF IRRIGATION, DRAINAGE AND FLOOD CONTROL IN NIGERIA FROM PRE-COLONIAL TIME TO 1999

E. U. Nwa
Professor of Irrigation Engineering and
Dean, Faculty of Engineering, University of Uyo, Nigeria.
Former Vice-President, International Commission on Irrigation and Drainage (ICID)

Spectrum Books Limited
Ibadan
Abuja • Benin City • Lagos • Owerri

Spectrum titles can be purchased on line at
www.spectrumbooksonline.com

Published by
Spectrum Books Limited
Spectrum House
Ring Road
PMB 5612
Ibadan, Nigeria.
e-mail: admin1@spectrumbooksonline.com

in association with
Safari Books (Export) Limited
1st Floor
17 Bond Street
St Helier
Jersey JE2 3NP
Channel Islands
United Kindom

Europe and USA Distributor
African Books Collective Ltd.
The Jam Factory
27, Park End Street
Oxford OX1, 1HU, UK

© E. U. NWA

First published, 2003

All rights reserved. This book is a copyright and so no part of it may be reproduced, stored in a retrieval system, or transmitted in any form or by any means, electronic, mechanical, electrostatic, magnetic tape, photocopying, recording or otherwise, without the prior written permission of the copyright owner.

ISBN: 978-029-402-3

Printed by Printmarks Ventures, Ososami, Ibadan

DEDICATION

To my wife, Ema, and children
Ini, Inemesit, Ido, Unwana,
Odudu and Kukpamba.

CONTENTS

LIST OF FIGURES

LIST OF TABLES

LIST OF APPENDICES

ACRONYMS AND ABBREVIATIONS

ABU	Ahmadu Bello University
ADP	Agricultural Development Project
AERLS	Agricultural Extension Research Liaison Service
AIRBDA	Anambra-Imo River Basin Development Authority
APMEPU	Agricultural Projects Monitoring, Evaluation and Planning Unit
APMEU	Agricultural Projects Monitoring and Evaluation Unit
ARCN	Agricultural Research Council of Nigeria
BASIRDA	Bauchi State Integrated Rural Development Authority
BCGA	British Cotton Growing Association
BIP	Bakolori Irrigation Project
BORBDA	Benin-Owena River Basin Development Authority
CBDA	Chad Basin Development Authority
CDC	Commonwealth Development Corporation
CRBDA	Cross River Basin Development Authority
ERLS	Extension Research Liaison Service
FACU	Federal Agricultural Coordinating Unit
FAO	Food and Agriculture Organization of the United Nations
FDA	Federal Department of Agriculture
FDID	Federal Department of Irrigation and Drainage
FDRD	Federal Department of Rural Development
FDWR	Federal Department of Water Resources
FMANR	Federal Ministry of Agriculture and Natural Resources
FMARD	Federal Ministry of Agriculture and Rural Development
FMAWRRD	Federal Ministry of Agriculture, Water Resources and Rural Development
FMWR	Federal Ministry of Water Resources
HJRBDA	Hadejia Jama'are River Basin Development Authority
IADP	Integrated Agricultural Development Project
IAR	Institute for Agricultural Research
IAR&T	Institute for Agricultural Research and Training
ICID	International Commission on Irrigation and Drainage
IITA	International Institute of Tropical Agriculture
JARDA	Jigawa Agricultural and Rural Development Authority
JICA	Japan International Cooperation Agency
KARDA	Kebbi Agricultural and Rural Development Authority
KNARDA	Kano Agricultural and Rural Development Authority

KRIP	Kano River Irrigation Project
LAIP	Lower Anambra Irrigation Project
LBRBDA	Lower Benue River Basin Development Authority
LCBC	Lake Chad Basin Commission
LCRI	Lake Chad Research Institute
LNRBDA	Lower Niger River Basin Development Authority
MANR	Ministry of Agriculture and Natural Resources
MNR	Ministry of Natural Resources
NAERLS	National Agricultural Extension Research Liaison Service
NBA	Niger Basin Authority
NBTE	National Board for Technical Education
NCRI	National Cereals Research Institute
NCST	National Council for Science and Technology
NCWR	National Council on Water Resources
NDA	Niger Dams Authority
NDBDA	Niger Delta Basin and Rural Development Authority
NEPA	National Electric Power Authority
NESCO	Nigerian Electrical Supply Company
NFDP	National Fadama Development Project
NGO	Non-Governmental Organization
NIFFR	National Institute for Freshwater Fisheries Research
NINCID	Nigerian National Committee on Irrigation and Drainage
NSC	Nigerian Sugar Company
NSCID	National Sub-Committee on Irrigation and Drainage
NSTDA	National Science and Technology Development Agency
NTCWR	National Technical Committee on Water Resources
NWRI	National Water Resources Institute
NWRMP	National Water Resources Master Plan
OORBDA	Ogun-Oshun River Basin Development Authority
RBRDA	River Basin and Rural Development Authority
RNC	River Niger Commission
SARDA	Sokoto Agricultural and Rural Development Authority
SCIP	South Chad Irrigation Project
SRRBDA	Sokoto-Rima River Basin Development Authority
SSC	Savannah Sugar Company
SWA	State Water Authority
TCPC	Technical Committee on Privatization and Commercialization
TVA	Tennessee Valley Authority

UBRBDA	Upper Benue River Basin Development Authority
UNDP	United Nations Development Programme
UNRBDA	Upper Niger River Basin Development Authority
USA	United States of America
USRI	Unilorin Sugar Research Institute
USSR	Union of Soviet Socialist Republics
WHO	World Health Organization
WRECA	Water Resources Engineering Construction Authority
WUA	Water Users' Association

FOREWORD

The history of irrigation, drainage and flood control in Nigeria cannot be compared to the developments along the Rivers Nile (Egypt), Yangtze and Hwang-Ho (China) and Ganges (India) in terms of age. Nevertheless, starting from about 700 AD when the shaduf was thought to have been introduced into what is now Northern Nigeria from Egypt, the people of Nigeria have managed in their own way to protect themselves against floods and to harness the waters around them for irrigation.

In addition to using the shaduf, the people of Northern Nigeria learnt how to divert water from streams into the fields to irrigate their crops long before the colonial period. The people of Sokoto were able to extend their knowledge of constructing small "tanks" for domestic and animal water supply to constructing small "dams" for irrigation purposes. Modern irrigation was introduced into Nigeria in 1926 when the first formal irrigation scheme the "Kware Irrigation Scheme" was opened in Sokoto.

The history of irrigation, drainage, and flood control in Nigeria cannot be separated from the history of agriculture in the country and, indeed, from the history of Nigeria itself. This book has incorporated the social and colonial history of Nigeria to make the subject matter interesting to anyone who has some interest in the history of Nigeria.

The author has succeeded in digging into our archives to bring out useful information that perhaps could have remained buried for a long time. It is interesting to note from the book that in 1925, debates about the following were already taking place:

- sound design and construction of irrigation schemes;
- irrigation research and extension as an integral part of any irrigation scheme;
- proper operation and maintenance as some of the key factors in successful irrigation schemes; and
- farmer participation as being indispensable for "sustainable" irrigation development.

This book is not only about the history of irrigation, drainage and flood control in Nigeria but it also contains lessons that Nigeria must learn if it ever hopes to produce enough food and fibre for its population and raw materials for its industries.

Dr. Bukar Shaib (CFR)

PREFACE

At its meeting held in Tehran, Iran in 1977, the International Executive Council of the International Commission on Irrigation and Drainage (ICID) approved and constituted a Committee on "History of Irrigation." Initially the Committee decided to cover six regions, namely:

1. The Nile Valley and the Delta (Egypt)
2. Euphrates and Tigris basins (Iraq, Syria and Turkey)
3. Indus basin (India and Pakistan)
4. North Africa (Algeria, Morocco and Tunisia)
5. Southern Europe (France, Greece, Italy, Portugal and Spain), and
6. North America (Canada, Mexico and USA).

However, in 1980 at its meeting in London, the International Executive Council of ICID decided to cover the history of irrigation of other regions as well. West Africa was one of the regions included and Nigeria was asked to coordinate the compilation of the history of irrigation of West Africa. The Nigerian National Committee of ICID, therefore, requested the author of this book to coordinate the work in West Africa.

During the ICID Congress in Grenoble, France in 1981, the delegates from the West African countries present met and decided that each country should set up a small group to work on the project and give the first report in 1982 during the Afro-Asian Regional Conference of ICID in Lagos. Unfortunately, only few members attended the meeting in Lagos and follow-up actions did not produce positive results and, consequently, it was decided that the work in Nigeria would continue. Even in the case of Nigeria, the work became very slow for financial and other reasons. However, it is satisfying to produce the History of Irrigation, Drainage and Flood Control in Nigeria in this volume.

Although Nigeria did not produce very old and elaborate hydraulic structures such as those found in the Nile and Euphrates - Tigris valleys, its people still struggled to control water for their benefit in the form of irrigation and flood control quite early in their history.

The history of irrigation of a predominantly agricultural society such as Nigeria cannot be entirely separated from the political and social history of the people. Therefore, anyone who reads this book will not fail to gather a fair amount of information about the history of Nigeria.

Chapter 1 gives the general background information about Nigeria. Such information includes the climate, water resources, population and the political development of the country. Pre-colonial development up to 1900 is treated in Chapter 2. The date 1900 is recognized as the beginning of effective

British administration in Nigeria and it appeared that the period 1906 to 1924 was used to prepare for the first formal irrigation scheme in Nigeria. This period is treated in Chapter 3 while the scheme itself, the Kware irrigation Scheme is treated in Chapter 4.

The period between 1925-1959 witnessed a rising interest in irrigation and saw the development of several medium-scale irrigation schemes. This was probably because the period included the depression and World War II years of 1929-1945. The above period is covered in Chapter 5. Chapter 6 covers the period of the oil boom which witnessed the development and construction of large-scale irrigation schemes. The evolution of the Federal Ministry of Water Resources and Irrigation Administration in Nigeria are dealt with in Chapter 7 while the concept and evolution of River Basin Development are covered in Chapter 8. The problems of floods and droughts, and salinity and water logging are treated in Chapters 9 and 10, respectively.

Irrigation Research, Extension and Training (Chapter 11) and Land Tenure, Water Laws, Rights and Customs (Chapter 12) are all very important components of Irrigation and Drainage. The International Organizations, Treaties and Agreements with respect to Nigeria are investigated in Chapter 13. Irrigation, Drainage and Flood Control are still growing concerns in Nigeria; their current problems are analysed in Chapter 14.

History of Irrigation Drainage and Flood Control in Nigeria, is not meant only to provide facts and information about these matters but also to provide lessons that would guide the development of irrigation, drainage and flood control in Nigeria for the future.

Prof. E. U. Nwa
University of Uyo
Uyo

ACKNOWLEDGEMENTS

Many people contributed information for this book. The author is particularly grateful to the following persons who spent time in the various archives in Nigeria searching for particular types of information, some of which are contained in this volume. These are Dr. P.O. Erhabor of the University of Agriculture, Makurdi, formerly of the Ahmadu Bello University, Zaria; Mr. R.A.K. Jimoh operated from the Federal Department of Water Resources (FDWR), Kaduna and was assisted by Messrs Moyo A. Oyetayo, Ferdinan O. Okeia and Kehinde A. Afolabi; Mr. C.U. Ekezie operated from FDWR, Enugu; Mr. D.A. Ahagbuje from FDWR, Ibadan; and Miss Nnennaya Achi from FDWR, Lagos. Engr. E. O. Okeke of the Federal Ministry of Water Resources supplied some information from time to time.

Special thanks are due to Dr. B. Martins, former Director of the Federal Department of Water Resources who encouraged this work, late Engr. N.O. Popoola, Permanent Secretary, Federal Ministry of Agriculture, Water Resources and Rural Development who gave the financial support for the initial work in the archives; Mr. M.S. Ayinmodu, former General Manager, Niger River Basin and Rural Development Authority, Ilorin, gave office facilities for the work from time to time.

The author and the publisher are grateful to the following for permission to use their maps and drawings: Anambra-Imo, Chad Basin, Hadejia-Jama'are and Sokoto-Rima Basin and Rural Development Authorities; Nigerian Sugar Company, Bacita, Kwara State; and Water Resources and Engineering Construction Company (WRECA), Kano State. The author is also grateful to late Engr. H.L. Musa, Director, National Centre for Agricultural Mechanization, Ilorin for permission to use the Centre's computers to type the work and to Mrs. Victoria E. Uwaifo of the same organization for typing the initial draft of the work. Special thanks are due to the Nigerian National Committee of the ICID and its chairman, Engr. I.K. Musa who is also the Federal Director of Irrigation and Drainage, for financing the publication of this book.

The manuscript was reviewed by both Engr. N.O. Popoola and Dr. Bukar Shaib, the former Permanent Secretary and the former Minister, respectively, who took charge of Water Resources at the federal level during the early years of the Federal Ministry of Water Resources. The author is truly grateful to both gentlemen. However, any errors in this book are entirely mine.

Chapter 1

Introduction

1.1 Location

The Federal Republic of Nigeria lies between latitudes 4° 10' and 13° 50' N and longitudes 2° 15' and 14° 45' E. The country is bordered in the west by the Republic of Benin, in the north by Niger, in the east by Chad and Cameroon, and in the south by the Gulf of Guinea where it has an 800 km coastline (Fig. 1.1).

1.2 Climate

1.2.1 *Rainfall*

The climate of Nigeria is governed largely by the seasonal movements of the Inter-Tropical Convergence Zone. The rainfall over the country is therefore, strongly seasonal. The annual rainfall varies from more than 4,000 mm in the coastal areas of the south east to below 250mm in the extreme north east. There is a regular decline in the annual rainfall from the south to the north; this regularity is often disturbed by the influence of the Jos plateau. The wet season usually occurs between July and September in the north and between April and November in the south.

1.2.2 *Evaporation*

The annual Class A pan evaporation varies from 1,500 - 2,000 mm in the south, 2,000 - 2,500 mm in the central region to 3,000 - 3,500 mm in the north.

1.2.3 *Temperature and Humidity*

In the north, the maximum mean temperature rises to 40°C between March and May while the mean minimum temperature drops to 12-18°C from November to January. Along the coast, thc highest monthly mean may be up to 32°C while the lowest mean minimum may be about 21°C. The mean daily relative humidity varies from 14 to 74 per cent in the north and from 71 to 79 per cent in the south.

Fig. 1.1: Map of Nigeria showing 36 states and Abuja

1.3 Population

Nigeria remains the most populous country in Africa. The population grew from 30.4 million in 1952 to 88,992,220 in 1991.

1.4 Administration

The summary of the changes in the administration of Nigeria is given as a brief history of Nigeria in Table 1.1.

Table 1.1. A brief history of Nigeria indicating the various administrative changes

S/No.	Year	Event
1.	1861	King Docemo of Lagos, the 13th in succession, was forced to cede his kingdom, and so Lagos became a British Colony.
2.	1874	The Colony of the Gold Coast was inaugurated; Lagos was included in this Colony.
3.	1884-5	Berlin conference allowed Britain to lay claims to the area now known as Nigeria.
4.	1885	The Oil Rivers Protectorate was established with its headquarters in Calabar.
5.	1886	Lagos was separated from the Gold Coast and established as an independent colony.
6.	1888	Yorubaland was co-opted into Lagos and the whole area was administrated as the Colony and Protectorate of Lagos.
7.	1893	The Oil Rivers Protectorate was extended inland to Leakage and Bene River and renamed the Niger Coast Protectorate.
8.	1900	a) Britain chose the name Nigeria (out of suggestions such as Niger Sudan, Negretia, Goldesia) for its territories in the area. b) The Niger Coast Protectorate was renamed the Protectorate of Southern Nigeria; its headquarters remained in Calabar. c) Britain declared a Protectorate of Northern Nigeria with its headquarters at Lokoja.

S/No.	Year	Event
		d) The whole of Nigeria was reorganized into three administrative areas under the colonial office, namely: i) The Colony and Protectorate of Lagos, ii) The Protectorate of Southern Nigeria, and iii) The Protectorate of Northern Nigeria.
9.	1906	The Colony and Protectorate of Lagos was merged with the Protectorate of Southern Nigeria as The Colony and Protectorate of Southern Nigeria, with Lagos as its headquarters.
10.	1914	a) The Northern and Southern Protectorates were amalgamated with its headquarters in Lagos. b) Nigeria had three groups of Provinces i) Colony of Lagos ii) Northern Province iii) Southern Province.
11.	1924	The mandated territory of the Cameroon (a German Colony before the 1st World War) was joined and administered with Nigeria.
12.	1939	Northern and Southern Provinces were broken into Eastern, Northern and Western Provinces.
13.	1951	The regional administration in each province was upgraded to a Regional Government.
14.	1954	A federal system of government was introduced. The federation had five parts: i) The Federal Territory of Lagos ii) Eastern Region iii) Northern Region iv) Western Region v) Southern Cameroon (a quasi Federal Territory).

S/No.	Year	Event
15.	1957	Regional self-government was granted to the Eastern and Western Regions. Southern Cameroon was administered as part of the Eastern Region.
16.	1959	a) Northern Region became self-governing b) Southern Cameroon left Nigeria.
17.	1960	Nigeria became an independent state under the British Commonwealth.
18.	1963	a) Nigeria became a Republic and member of the Commonwealth b) Mid-West Region was created from Western Region, therefore Nigeria had four Regions: i) East ii) Mid-West iii) North iv) West.
19.	1966	First coup *d' etat* in Nigeria.
20.	1967	a) The Nigerian Civil War started b) 12 states created in Nigeria out of the four Regions.
21.	1967	The Nigerian Civil War ended.
22.	1976	19 States created in Nigeria out of the 12 states. Abuja was chosen as the Federal Capital Territory.
23.	1987	Nigeria had 21 states
24.	1991	a) Nigeria had 30 states. b) The Federal Capital was moved from Lagos to Abuja.
25.	1996	36 states created in Nigeria out of the 30 states.

1.5 Land Use

A summary of land use in Nigeria is given as follows:

i.	Total land area of Nigeria	-	92,377,000 ha
ii.	Total cultivable area	-	71,000,000 ha
iii.	Total cultivated land	-	39,200,000 ha
iv.	Grassland	-	28,110,000 ha
v.	Forest land including woodland, tropical rain forest, mangrove swamps	-	13,680,000 ha
vi.	Wetland	-	2,510,000 ha
vii.	Others (including urban disturbed, bareland and water)	-	8,850,000 ha

1.6 General Geology

Nigeria is dominated by two broad geological units, namely, the crystalline rocks of the basement complex and younger sedimentary formulations. Approximately 50 per cent of the country is underlain by the basement complex rocks while the remaining 50 per cent is underlain by sedimentary formulations.

1.7 Water Resources

1.7.1 *Surface Water Resources*

There are four principal surface water catchments in Nigeria, namely, the Niger, Lake Chad, eastern littoral and the western littoral. The total annual runoff from all sources was estimated to be 257.30 x 10^9 m^3 (NWRMP,1995).

a) Niger

The Niger river, with its left bank tributary, the Benue, drains 63 per cent of the country and forms an extensive delta at its mouth. The major tributaries of the Niger are the Benue (with its own tributaries of Katsina-Ala, Donga, Taraba, Gongola, Ankwe and Mada), Sokoto-Rima, Kaduna, Gurara, Gbako, Kampe and Kontagora.

b) Chad Basin

Approximately 95 per cent of the total river inflow into Lake Chad comes from the Chari-Logone river system which is entirely outside Nigeria. The main rivers that flow into Lake Chad from Nigeria are the Komadougou-Yobe, Yedseram and Ebeji.

c) *Eastern Littoral*

The Cross is the major river in the eastern littoral. The others are Imo and a smaller river, Qua Iboe. All these rivers flow into the Gulf of Guinea.

d) *Western Littoral*

A number of smaller catchments which drain the south-west of the country between the Niger catchment and Atlantic Ocean make up the western littoral. The most important of these rivers are the Yewa, Ogun, Oshun, Ona, Shasha, Omi, Oluwa, Owena, Osse, Ossiomo, and Benin.

1.7.2 *Groundwater Resources*

A reliable estimation of the total groundwater resources of Nigeria is not yet available. The National Water Resources Master-Plan puts the potential groundwater yield at 51.93 x 10^9 m^3 (NWRMP,1995).

Chapter 2

Pre-Colonial Period up to 1900

2.1 Background

From pre-history to the British colonial period, the people who inhabited the area now known as Nigeria lived in independent communities. Some of these communities later developed into empires, states and kingdoms.

In the drier north, the Kanem empire developed in about AD 700, reached its zenith from 1221 to 1259 and collapsed between 1259 and 1472. Borno succeeded Kanem in 1472, reached the peak of its power between 1272 and 1603 and then collapsed in 1846. Also developing in the north were the Hausa states represented by such states as Kano (AD 999 - AD 1706) and Katsina which took over from Kano and collapsed in 1807. Other powers in the north included Kebbi, Nupe and Jukun empires.

In the wetter parts of the area, the Yoruba empire was founded in the 14th century, reached its peak in the 18th century and collapsed in 1896. The great empire of Benin, founded in about 12th century reached its greatest splendour in the 15th and 16th centuries and then fell to the British in 1897. In this part also were the City-States of Nembe, Kalabari, Bonny and Old Calabar, the kingdom of Warri, and the village democracies of the Ibo and Ibibio (Stride and Ifeka, 1971).

Portuguese ships first reconnoitred the coasts of this area between 1469 and 1475. By 1510, the Portuguese had started the trans-atlantic slave trade. After 1713, Britain and France took over the control of the slave trade on the West African coast from the Portuguese and the Dutch.

The founding in Britain in 1788 of the African Association led to the discovery of the source of the Niger and the exploration of the interior region. The exploration of the interior resulted in the scramble for Africa by European powers and the Berlin conference of 1884-5 which allowed Britain to lay claims to the area now known as Nigeria.

Before the Berlin conference, Britain had carried on a great deal of trading and political activities in the region. For example, in 1849, John Beecroft was appointed the first British Consul for the Bights of Benin and Biafra to supervise British trading interests in the area. Beecroft laid the foundation of the British power in Nigeria. In 1861,

Docemo, the king of Lagos, was forced to cede his kingdom, and Lagos became a British colony.

The Oil Rivers Protectorate was established in 1885 with its headquarters in Calabar. This protectorate was extended inland to Lokoja and Benue river in 1893 and renamed the Niger Coast Protectorate. The protectorate was expanded to include Benin Empire following the British occupation of Benin in 1897. Meanwhile, Yorubaland had been co-opted into Lagos in 1888 and the whole area was administered as the Colony and Protectorate of Lagos.

In 1900, the British Government chose the name Nigeria (out of suggestions such as Niger Sudan, Negretia and Goldesia) for its territories in the area. In the same year, the Niger Coast Protectorate was renamed the Protectorate of Southern Nigeria; a High Commissioner was appointed for the protectorate and its headquarters remained in Calabar. Also, in the same year, Britain declared a Protectorate of Northern Nigeria with headquarters at Lokoja and Frederick Lugard as its first High Commissioner. In 1900, the whole of Nigeria was, therefore, reorganized into three administrative areas under the Colonial Office, namely, the Protectorate of Southern Nigeria, the Protectorate of Northern Nigeria and Colony and Protectorate of Lagos.

In 1906, the Colony and Protectorate of Lagos were merged with the Protectorate of Southern Nigeria as the Colony and Protectorate of Southern Nigeria with Lagos as its headquarters. Also, in 1906, Lugard completed his conquest of Northern Nigeria. This date is, therefore, recognized as the beginning of effective British administration in Nigeria (Crowder, 1966).

2.2 Shaduf

The shaduf (*jigo* in Hausa)[1] is a simple device for lifting water using the principle of the lever. A bucket or calabash is tied to a rod suspended from one end of a long pole. The other end of the pole carries a counterweight. The fulcrum and the support system are provided by a Y-shaped tree branch. By pulling the rod down, the operator lowers the calabash into the well, pond or stream where it is filled with water. When the rod is released, the counterweight lifts the calabash to the level of the operator where the water is poured into an inlet from which

[1] *Variously known as shadoof, shadouf, chadouf, dhenkali, dhenkli, dhingli, khetara, kerkaz, kheeraza, picottah, picotas, sweep, swape, lat, guenina, bascule, and counterpoise lift (Wood, 1976).*

it flows into a canal leading to the field to be irrigated.

The Y-shaped support and fulcrum system is about 1.5m high and 10cm in diameter. The length of the pole which is laid across the Y-frame varies from 30 to 100 cm and the diameter is 12 to 15 cm. The length of the rod that is tied to one end of the pole varies from 1 to 7m depending on the height of the device from the water surface. The counterweight weighs between 9 and 55 kg and is made of dried mud bonded with grass. Most of the shadufs used in Nigeria can lift water from one up to three metres. The lifting rate varies from 0.12 to 5.22 m^3/hr depending on the age, strength and experience of the operator and the time of day (Nwa, 1981; Erhabor, 1982).

The shaduf system is mostly used in the *fadama*. *Fadama* is a Hausa word applied in Northern Nigeria to low-lying flood plains periodically subject to flooding and often with a very high water table. This land is capable of supporting crop growth throughout the year except that some portions may be waterlogged or flooded during the rainy season. During the dry season, intensive vegetable production is common on the fadama land. Tomatoes, onions, potatoes, peppers, garden eggs and carrots predominate. Individual holdings of the shaduf farmers are usually less than 0.4 ha per farm family (Nwa, 1981). The components and the operation of the shaduf are illustrated in Figure 2.1.

a. Getting water out of the stream

b. Pouring water onto the shaduf bed

Fig. 2.1: Components and operations of the shaduf

The shaduf is said to have originated from the Nile valley in Egypt. They were first recorded in tomb drawings dating from 2000 BC (Smith, 1975). These same shadufs have been operating in Northern Nigeria for centuries. The question is, how and when did the device move from Egypt into Nigeria? Or in fact, did it move into Nigeria directly from Egypt or from a third region or country?

Ancient Ghana could be said to have started as an empire in about AD 300. From this time, trade gradually developed between Ghana and North Africa. Ghana empire, and trade between Ghana and North Africa, reached their peak in AD 990. By 1068 Ghana had become the most powerful and the largest state in West Africa (Stride and Ifeka, 1971). These long years of contact with North Africa could have brought the shaduf to Ghana. However, Ghana was too remote from Hausaland in Northern Nigeria and any contact and influences, if any, was insignificant.

Mali empire emerged about 1230 AD after the collapse of Ghana. Between 1312 and 1337 AD, the rulers of Mali had built an empire that became known not only in the Islamic world but also in Europe. Timbuktu had become one of the foremost centres of Islamic scholarship, attracting Moslem scholars from all over the Islamic world. The rulers of Mali during this period, developed important relationships with the rulers of North Africa such as the Sultan of Fez. This was certainly another opportunity for the shaduf to reach Mali if it had not already reached through Ghana.

Songhai empire is said to have started in about AD 690 or 850 when invaders from "north east" who were probably the Zaghawa nomads from the north of Lake Chad, established a new ruling class in Songhai. In either AD 1009 or 1100, the capital of Songhai was moved to Gao, and the city gradually became the leading centre of trade in West Africa. Merchants were attracted from Morocco and Tunis. It was Songhai that brought the Mali empire to an end. Songhai then became the centre of trade and learning with frequent exchanges of ideas with Fez, Tunis and Cairo. Again, the shaduf could have moved into the area from Mali or directly from North Africa.

In about AD 700, groups of black nomads known as Zaghawa, arrived in the Lake Chad region. Forced towards the lake by both the shifting desert and the Arab invasions from North Africa, they settled in the lake region and thus the Zaghawa empire emerged. Some of the nomads responded to the favourable lake area by joining the local people in settled agriculture (Crowder, 1966; Stride and Ifeka, 1971). Zaghawa kingdom was the predecessor of Kanem-Borno empire. From 1085-97 under the rule of Mai Ume, the Kanem-Borno empire had become firmly established and considerable trade contact had been made with Tunis, Tripoli, Fezzan and Cairo. It would be difficult to believe that a partly agricultural empire such as Kanem-Borno would fail to make contact with the shaduf in North Africa.

The Hausa states did not develop until immigrants arrived from Kanem-Borno empire. For example, Kano was still a little village until the Habe rule which started in AD 999. The situation was not different in Katsina. Kano and Katsina did not attract North African merchants until 1438-1452. But before this time, the two states had considerable contact with Mali, Songhai and Kanem-Borno empires. Since agriculture was the main source of wealth of Kano and Katsina, it would not be surprising if the two states showed early interest in the shaduf which was certain to boost their agricultural production.

The above discussion and the map in Figure 2.2 show that the shaduf could have entered Northern Nigeria through three possible routes. These are Ghana-Mali-Songhai route, Kanem-Borno route or direct North African route. Although Songhai defeated Kano and Katsina in 1512 and made them part of Songhai empire, Kanem-Borno had more influence on Kano and Katsina and consistently exacted tribute from them. Kano had opened up direct trade with Kanem-Borno between 1438-1452. If the fact that the rulers of Hausaland are said to have migrated from Kanem-Borno is considered along with the relatively shorter distance between Hausaland and Kanem-Borno vis-a-vis Hausaland and Songhai empire, the movement of the shaduf from Kanem-Borno to Hausaland appears to be more likely.

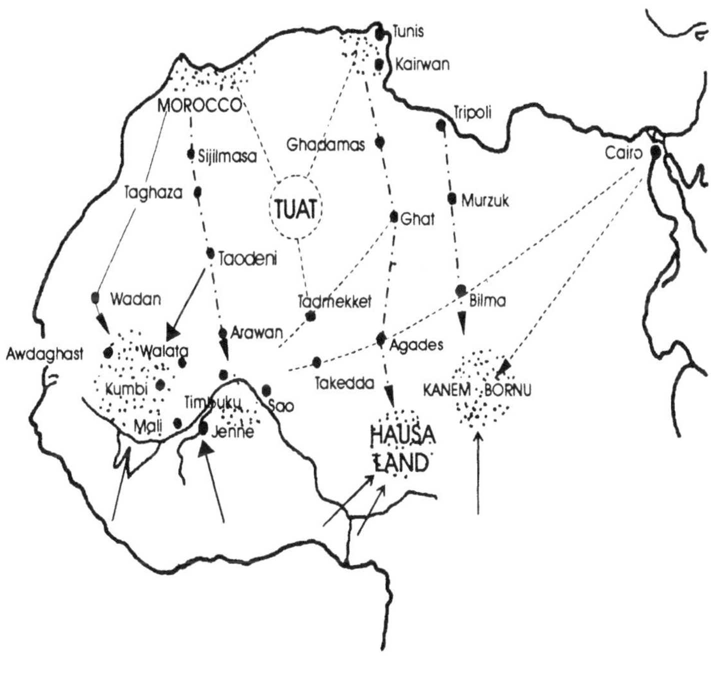

Fig. 2.2: The Trans-Sahara trade route (Stride and Ifeka, 1971)

In conclusion, it may be said that the shaduf probably first moved into the north eastern part of Nigeria (Kanem-Borno) from North Africa (Tripoli - Cairo) after AD 700. The shaduf may have spread into Hausaland from either Mali-Songhai empire or from Kanem-Borno empire after AD 999.

2.3 Sub-Irrigation

An irrigation method whereby artificial or natural groundwater is manipulated for the benefit of the crop is known as sub-irrigation. Two sub-irrigation systems are recognized:

1. *Continuous Flow System*

   In this system, open ditches 30 to 100 cm deep, with spacings of 15 to 30cm, are kept flowing with water which seeps into the subsoil and then reaches the plant roots by capillary action.

2. *Intermittent Flow System*

   In this system, open drainage ditches or drainage pipes are installed in the area to be irrigated. To irrigate the field, the drains are blocked. This causes the water table to rise to a level that allows water to get to the plant roots by capillary action.

It is obvious that the sub-irrigation method can be used only in restricted circumstances. The topography must be smooth and nearly level, there must be an impermeable layer at a reasonable depth, the surface soil must be highly permeable loam or sandy loam, and the water supply must be abundant (Zimmerman, 1966; Michael, 1978).

Sub-irrigation has been practised in the north eastern part of Nigeria for centuries. It is said to have been introduced long before 1775 most probably from Egypt. The system is concentrated around the villages of Marma, Kaban, and Guri in the Burum Gana and Marma flood plains south of Nguru.

The Burum Gana and Marma rivers have natural commands over their flood plains in the area. The flood plains slope away from the rivers on either side. Long irrigation ditches are dug at right angles to the rivers, and irrigation water seeps from the ditches into the surrounding soils to irrigate the crops grown. Drainage channels are constructed to drain the land, and dykes are built to offer flood protection to the irrigated areas. These channels and dykes are maintained annually through massive communal effort. In 1975, 6 to 10,000 ha of land was under irrigation in the area. The major crops grown were cotton and rice. Other crops included sugar cane, tobacco, and vegetables (Schultz International Limited, 1975).

2.4 Other Systems of Irrigation

Dates as early as BC 5000 and from BC 2800 to 1500 have been suggested for the development of agriculture in West Africa. Some of the crops domesticated in the region include fonio or hungry rice *(Digitaria exilis)*, African rice *(Oryza glaberrina)*, okra *(Hibiscus esculentus)*, and water melon *(Citrullus vulgaris)* (Havinden, 1975). These are not the most important of the indigenous crops, but they are some of the crops generally grown under irrigation in the drier parts of Northern Nigeria.

In about AD 500-800, crops such as banana *(Musa sapientum)*, plantain *(Musa paradisiaca)*, Asiatic rice *(Oryza sativa)*, and sugar cane *(Sacchurum officinarum)* were introduced from South-East Asia (Havindin, 1975). Among the later introductions from South America through Portuguese contacts with the Guinea Coast were maize *(Zea mays)* and tomato *(Lycopersicon lycopersicum)* (Agboola, 1979).

The above crops, among others, have been grown in the fadama lands of Northern Nigeria from the time of their introduction to date. Various methods were used to irrigate these crops. The most widespread was the use of the natural flooding of rivers and lakes. The most important river systems and lakes in this respect were the Sokoto-Rima, Kano-Hadejia-Yobe, Yedseram, Niger, and Lake Chad. For example, by the time the colonialists got to Argungu and Birnin Kebbi Divisions of the Sokoto province in the 18th century they found that the people had been growing rice by the above method in water which varied in depth from a few centimetres to more than 2 metres (KMAG, File No. 11019 Vol. I, 1924).

The diversion of streams for irrigation was also noted to have been practised by the local people before the colonial period (KMAG, File No. 26902, 1919). In about 1880, canals were constructed by the local people in Edozhigi in Niger province which enabled them to divert water from streams to irrigate their rice fields. The people of Kware north of Sokoto were also excellent farmers who constructed canals to irrigate their crops of rice, sweet potatoes and cassava.

The people of Tsamia, Kamaru, Rabah, and other areas of Sokoto province irrigated their rice fields by constructing water conservation works. These works impounded water at high floods and retained it after the river level had fallen. This often resulted in large flat areas being flooded up to about 60 cm deep for rice cultivation (KMAG, File No. 11019, 1919).

Before the advent of dam construction the local people in Sokoto province had been constructing small tanks for domestic water supply and for their cattle. These tanks later encouraged the people to construct dams for irrigation purposes. One such tank was constructed in about 1900 in Boho valley near Wababi in the Dange-Wababi area of Sokoto Division by Sarkin Rafi the District Head of the area at the time. The local people believed that it was in the location of the tank that Sarkin Musulmi Bello, years before the tank was built, found water for his army when they encamped at Wababi. The water in the tank was maintained by surface runoff and by natural springs at the bottom of the tank.

After re-excavation in later years, the tank was able to supply enough water for the village people and for 300 to 400 cattle during the dry season (SOKPROF, File No. 6613, 1950).

Chapter 3

The colonial period between 1900 and 1906

3.1 General

The period between 1900 and 1924 included the foundation years of the colonial regime (1900 -1912) when there was bias for peasant farming and emphasis was more on forestry and mining. However, the famine in Northern Nigeria during this period must have directed attention to irrigation as one of the means of increasing food production in the region. Between 1902 - 1914 there were patchy, moderate to severe famines in various parts of Northern Nigeria.

Probably the most severe famine in Northern Nigeria since 1855 was said to have occurred between 1912 - 1914 (Tarhule and Woo, 1997). This was followed by the first World War (1914 - 1918). As stated earlier, these events must have stimulated the desire to use irrigation to solve the food problems of the drier north in particular and Nigeria in general.

3.2 Notable Events

E. Percy C. Girouard became the Governor of the Protectorate of Northern Nigeria in 1907 after Frederick Lugard who was High Commissioner between 1900 and 1907. Being based in Zungeru on the Kaduna river, and from his travels and reports, he became impressed by the agricultural and irrigation potentials of most of Northern Nigeria. He was particularly impressed by the flood plains of the Niger between Baro and Jebba which he considered to be "possibly the richest land of all northern Nigeria." He was aware of the fact that the local population had been growing, under irrigation, crops such as rice, onion, sugar cane and tobacco. He, therefore, decided that experiments should be carried out to test the performance of these and other crops under irrigation.

An experimental irrigation farm, the "Baro Experimental Irrigation Farm" was, therefore, started at Baro in November 1908. The objective of this farm was to test the performance of certain crops during the dry season under irrigation. A diaphragm pump was used to lift water from the river to irrigate the farm. When the pump was not available, the shaduf was used for the irrigation. The work on the farm was carried out entirely by prison labour.

The crops experimented upon were onion, cotton, tobacco and groundnut. The yield of seed cotton was said to be satisfactory. It was reported that the groundnut could have yielded better if it had been planted on ridges as was the local practice instead of planting it on the flat. The sizes of the onion were said to be from "moderately large to undersize". The tobacco crop was fairly successful although it was attacked by insects. Samples of all the crops were sent to the Imperial Institute in Britain for analysis (KMAG, File No. 41577, 1909).

Other colonial officers such as residents and agricultural staff were keen to see scientific irrigation introduced into Northern Nigeria. One such officer was P.H. Lamb who was Director of Agriculture, Northern provinces in 1919. In that same year he sent a letter to the Secretary, Northern provinces, Kaduna, asking him to transmit his views on irrigation in the Northern provinces to the governor. In his letter he listed the following as the principal methods adopted for irrigation:

a) The diversion of streams
b) Dams for raising the level of streams
c) Storage reservoirs
d) Raising water by hand or by mechanical means.

He noted that method (a) and (d) are generally employed by the natives in Nigeria. He wanted (b) and (c) introduced into Nigeria.

However, Lamb was reprimanded for sending a letter straight to the Governor instead of passing it through the Deputy Governor. Records have not indicated whether the letter eventually got to the Governor or not (KMAG, File No. 26902, 1919).

In 1919 the Secretary, Northern provinces, sent a circular to all Residents, Northern provinces. The content of the circular was as follows (KMAG, File No. 11019, Vol I, 1919):

> I am directed to draw your attention to the necessity of increasing the cultivation of rice for the following reasons: In 1914, no less than 212.000cwt (10.770 tonnes) of rice was imported into Nigeria. At the present moment, there is a danger of a serious world shortage of rice. It is therefore of great importance that in this respect Nigeria should become self-supporting. I am to ask your views as to the measures that might be taken in your province to stimulate the production of rice.

All the provinces replied that if supplied with the better Ceylon swamp rice variety and if the producer prices were improved they would produce enough rice for the province and more for export. Ilorin province added that in 1917 and 1918, they exported from the Province 215 tonnes and 332 tonnes, respectively, by railway alone. This did not include the "very large quantity" also exported by river.

Of all the provinces, Sokoto province gave the most comprehensive response. They noted that rice was a favourable crop with the people and large quantities were already being grown. However, the price structure needed to be changed. The local price was only "a penny a pound."

At that time, the Western railway line which started at Ebute Metta, Lagos in 1896 had reached Ibadan in 1900 and Jebba in 1909; the Northern line started from Baro in 1907 and reached Kano in 1911; and the Jebba-Minna line reached Minna in 1913. Sokoto province believed that if this railway system could be extended to Sokoto, the transport system could enable rice production to be greatly expanded. Most of all, it was indicated that rice cultivation would be greatly increased if the existing water conservation works were made more secure and if modern irrigation facilities were provided.

In 1913, the safe average yield of rice in the district of Silame was given as 2690kg/ha. In the same year, it was reported that the yield of rice in Argungu District varied with the depth of water as given in Table 3.1.

Table 3.1: Yield of rice in Argungu District, 1913

Depth of water less than given number, m	Yield of rice kg/ha
0.30	392
0.60	1177
0.90	2354
1.20	3531
1.50	4708
1.80	5885
1.80	7061

Source: KMAG, File No. 11019, 1919.

The report from Sokoto province contributed to the increase in irrigation activities in the province shortly after 1919. The provinces which responded were Bauchi, Borno, Ilorin, Kano, Kontagora (Provincial Office), Muri, Nassarawa, Nupe, Sokoto, Yola and Zaria.

3.3 Edwardes' Dams, Canals and Weirs

H.S.W. Edwardes probably worked for some time in Sokoto province before he became the Resident of the province in 1921. He was succeeded as the Resident of the province by G.W. Webster in 1922 while he moved to Benue province and became Resident there in 1923. While Edwardes was in Sokoto province, he suggested that gauge readings of the flow of Sokoto river at Sokoto be kept and he also supervised the construction of a large number of small dams and canals in the province. Some of these included a dam near Raba to increase rice cultivation in the area, a canal upstream of a tapki (cattle watering point) near Sheyi, and a dam and canal near the village of Bakanwa. Others included dams near Yartsokua and Tumba, a dam near Dampo, three dams near Rimi and two canals near Dankaiwa. The locations of these projects are shown in Figure 3.1.

The most important of Edwardes' structures was the dam built across Bobo river, a tributary of Sokoto river, in 1918. Lake Natu is an extensive lake formed just above the junction of Bobo river with Sokoto river. The dam built across the outfall of this lake increased the water level in the lake by about 1.5m. But owing to the very flat nature of the area, the land inundated became very extensive and crops were damaged up to the villages of Gumji and Yarkofoji (Fig. 3.1). The local people complained about this damage and in 1922 the dam was cut open to reduce the reservoir area.

Edwardes' structures were constructed not only upstream from Sokoto town but also downstream and in the Western uplands. For example, a small dam was built by Edwardes just above the village of Binji, north of Sokoto, for stock watering.

Two experimental weirs were built by Edwardes in 1921, one about half way between Tozo and Silame (both are west of Sokoto) and the other near Silame. If these weirs were to withstand the floods, then complete weirs were to be built across the valleys at Tozo, Gwamatsi, and Wamako.

Rainfall records were said to have been kept in Sokoto since 1904 and in 1922, the 19-year mean annual rainfall was said to be 678 mm.

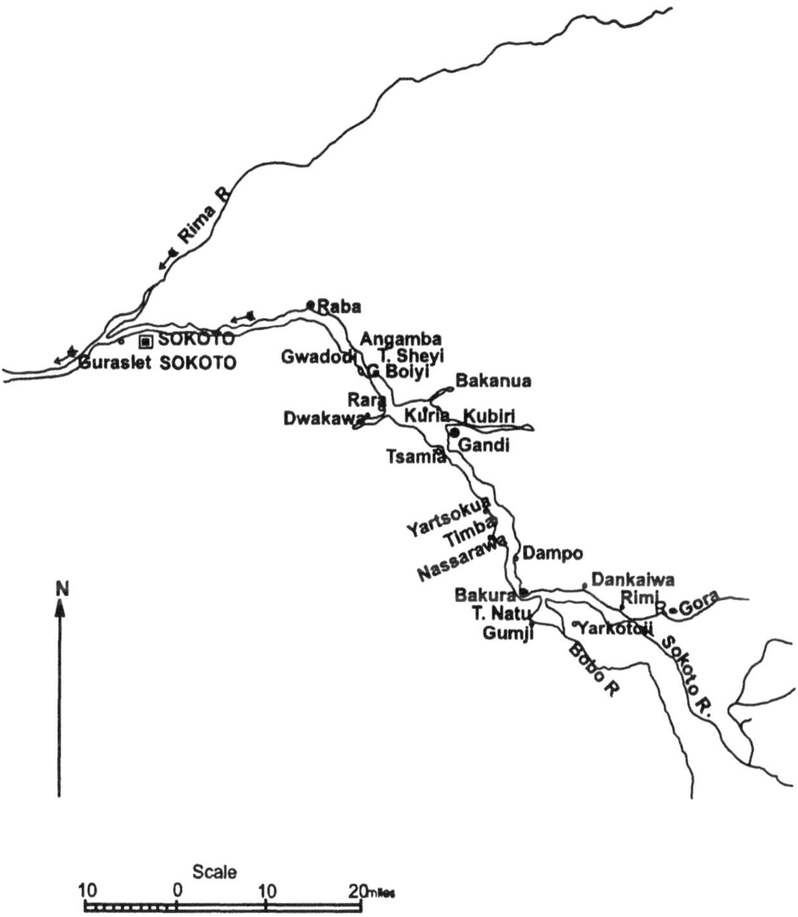

Fig. 3.1: Location of Edwardes' project in Sokoto province

However, in 1922 a total of 899 mm of rainfall was recorded most of which fell in April. This resulted in an abnormally high flood in Sokoto province. Certain villages such as U. Lalle and part of Dan Gerko were washed away. Also, as a result of the flood, the Tsamia and Dampo dams were washed away, the Yartsokua, Tumba, and Rafin Caro dams were breached. Although the Silame experimental weir remained intact, the Tozo weir was washed away. The Bobo dam had to be cut open (KMAG, File No. 10708, 1923).

Since Edwardes was not an engineer and did not probably have previous experience with water conservation structures, he did not provide spillways and good foundation for his structures. The Secretary of Northern provinces considered these schemes to be "amateur" and haphazard. Also because of the damage done and the complaints of the people, the Secretary of the Northern provinces invited in 1923, Col. M.R. Collins, an irrigation engineer, to visit Sokoto and investigate "the possibility of water control for rice growing in the valley of the Sokoto river."

3.4 Colonel Collins' visit to Sokoto Province

Colonel Collins arrived in Sokoto in September 1923 and explored the province following the major rivers until June 1924. The exploration started with Sokoto river upstream of Sokoto up to Bakura and then downstream as far as Raha towards the confluence of Sokoto and Niger rivers.

He inspected Bunsuru and Gagare rivers which combine to form Gulbin Kamwa (local name) as well as Rima river; Gagare, Bunsuru and Dan Gulbi combine to form Rima. He also explored almost the entire length of Zamfara or Gindi river, Gaminda, Gavan Gulbe, and Bobo rivers as well as small streams such as Dandaye and Shalla. The existing dams, canals, tapkis and lakes were inspected and possible dam sites and irrigable fadamas were noted. Figure 3.2 shows the map of Sokoto province emphasizing the areas covered by Colonel Collins.

Collins (KMAG, File No. 10708, 1923) was highly impressed by Lake Natu and the surrounding country and, therefore, he stated that "Lake Natu appears to offer from the technical standpoint the most promising point from which to begin irrigation development on the Sokoto river."

Collins observed the various crops grown in various parts of the province. Cotton was the favourite crop in the districts of Sabon Birni

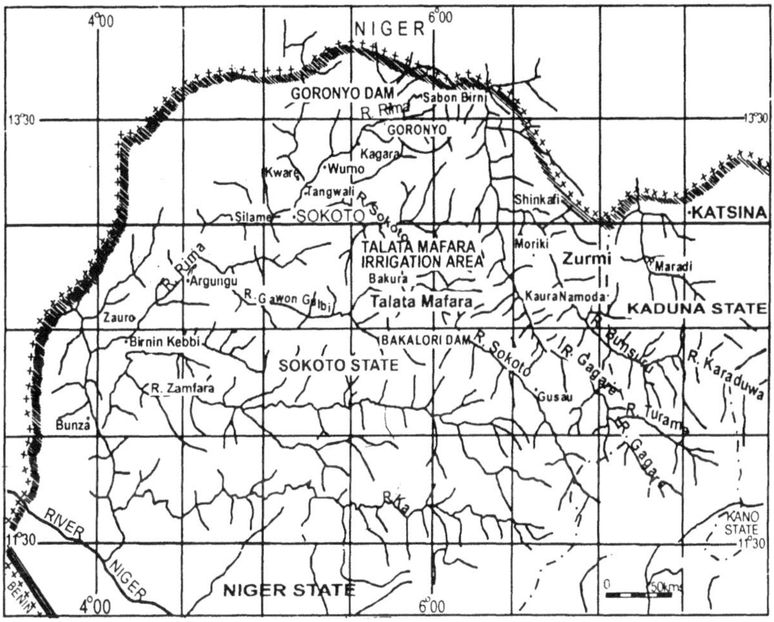

Fig. 3.2: Map of Sokoto province showing the rivers and areas explored by Col. M. R. Collins

and Isa in the fadamas of Bunsuru and Gagare. Boding district and Koffe in the district of Gwadabawa were noted for wheat cultivation.

Rice was the staple crop grown in the Emirates of Argungu and Gwandu. In 1924, Collins estimated that about 101,000 ha of excellent fadama land was available for rice cultivation in the area. About 16,200 ha of rice was actually grown in Argungu. The local people informed Collins that they could identify up to 30 varieties of rice grown in their district.

In 1923 about 40 to 50 per cent of the available rice land between Silame and Argungu remained uncultivated. The explanation Collins obtained for this was that the opposing Emirates of Sokoto and Argungu were in the habit of mutually raiding one another before the British occupation of the area. In the course of their raids, the Sokoto people were able to obtain large numbers of slaves who were used for extensive rice cultivation. After the British occupation this labour force dried up and, therefore, the area cultivated for rice was reduced.

Other crops under cultivation in Sokoto province and observed by Collins included guinea corn, millet, sweet potato, cassava, maize, tobacco, indigo, bean, and onion.

In his report after the visit, Collins stressed the importance of a contoured survey of any area in which an irrigation project was to be sited. He suggested that a gauge should be installed below the junction of the Sokoto and Rima Rivers. He also stressed the importance of establishing a government experimental farm placed in an irrigated area to demonstrate and teach irrigation technology to the farmers.

Collins recommended that four specific projects should be carried out. These were:

1. Shalla canal project and experimental farm near Kware
2. Kandole canal project near Dandaye
3. Girafshi canal project near Sokoto, and
4. A small inundation canal near Sokoto.

In July 1924 the Acting Secretary, Northern provinces, addressed a letter from Kaduna to the Chief Secretary in Lagos informing him that the four projects recommended by Collins had been approved for execution. A survey of Sokoto river downstream of Sokoto was to be carried out. These surveys were to include the sites of the four projects. An immediate appointment of an irrigation engineer for the projects was also approved (KMAG, File No. 10708, 1924).

Chapter 4

Kware Irrigation Scheme, 1925 – 1963

4.1 General

After his visit to Sokoto province, Colonel M.R. Collins recommended four irrigation projects for implementation (Section 3.4). However, it was the *Kware Irrigation Scheme* — the "Shalla Project", that received the strongest support from Mr. G.W. Webster, the Senior Resident of Sokoto province, and, therefore, this project was to be constructed first.

The scheme was situated on the northern side of the valley containing the Rima river starting from the junction of this river with the Shalla stream. It was about 1.6 km to Kware town and was on latitude 13° 13'N, longitude 5° 15'E and on an altitude of about 354 m above mean sea level.

The Shalla stream is a continuous "lake" of water about 90m to 180m wide and 11km long. It is a perennial stream that rises from the south-west of Gwadabawa-Kware plateau and is fed at the upper end by a number of perennial and non-perennial streams. The Shalla flows into Rima river near the town of Kware. The discharge of Shalla taken in November, 1923 was about 0.57 m^3/s.

Some of the reasons given to justify the Shalla project were the following (KMAG, File No. 30867, 1925):

- the Shalla is a perennial stream,
- from irrigation and engineering standpoint, the conditions of the site are most favourable and easy for an irrigation project,
- there will be no construction difficulty,
- politically, the people will be favourably disposed towards it,
- there will be no serious difficulty in getting the necessary land for the works,
- on the agricultural side it has the advantage that the people are already familiar with irrigation, and
- the farm would be used for experiments on irrigated agriculture and to teach people how to manage Shalla and other irrigation schemes.

4.2 Technical Details and Management

4.2.1 Planning

Colonel Collins proposed the Kware Irrigation Scheme in May 1925 and approval of the project was given in August of the same year. The scheme was to cover:

a) Drainage and land reclamation,
b) Irrigation, and
c) Flood control.

a) Drainage and Land Reclamation

The course of the Shalla stream between Kware and Tungwale, a distance of about 8 km, was often obstructed thus causing the water to spread into a series of mashes. The land thus inundated was said to be good agricultural land. The drainage and land reclamation project, therefore, was to:

i) canalize this portion of the Shalla, and
ii) construct two branch drains from the irrigated area back into the Shalla.

b) Irrigation

The Shalla lakes were 3m above the general level of the Shalla/ Rima river valley. A canal was to be constructed to convey water from the Shalla lakes and the upper part of the Shalla river to the fadama land to be irrigated. At intervals along the canal, earth ware pipe outlets were to be provided so that farmers could convey water to irrigate their farms as they needed it. The pipes could be closed when irrigation water was no longer required.

The irrigable area of about 16 square kilometres was a fadama land which lay between the angle formed by the junction of the Shalla stream with the Rima river (Brown, 1928).

c) Flood Control

The area to be irrigated was subject to flooding by water from the Rima river. To protect this land it was necessary to construct a flood control embankment starting from a location near the village of Marabawa. This embankment became known as the "Marabawa Flood Embankment." The embankment was to serve a dual purpose — to protect the irrigated land from flooding, and to serve as the key of an inundation scheme in which controlled and regulated flooding could be used for rice production.

4.2.2 Design Details

The contour map of the project area, the location of the irrigation canal and the Marabawa Flood Embankment, and other details of the project are shown in Figure 4.1. Figure 4.2 is a sketch giving more details of the canal and embankment system. "A" is the intake at the head of the irrigation canal; water was admitted through two 35.56cm (14") diameter pipes by means of two screw valves. "B" is a spillway which allowed flood water from the Shalla lake during the peaks of the rains to be drained through the irrigation and drainage canals into the fadama land outside the irrigated area.

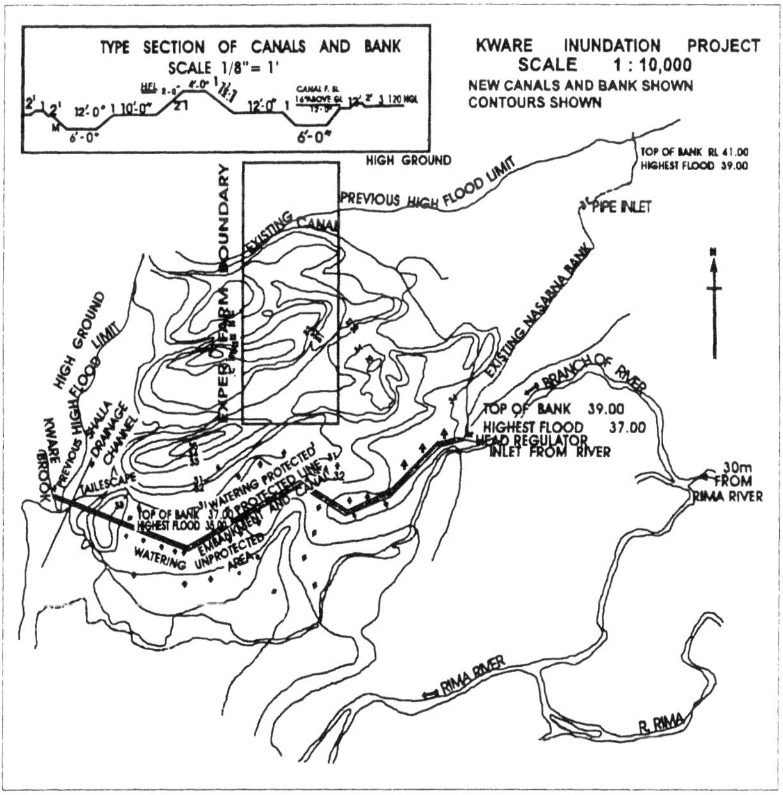

Fig. 4.1: Contour map and other details of the Kware Irrigation Project

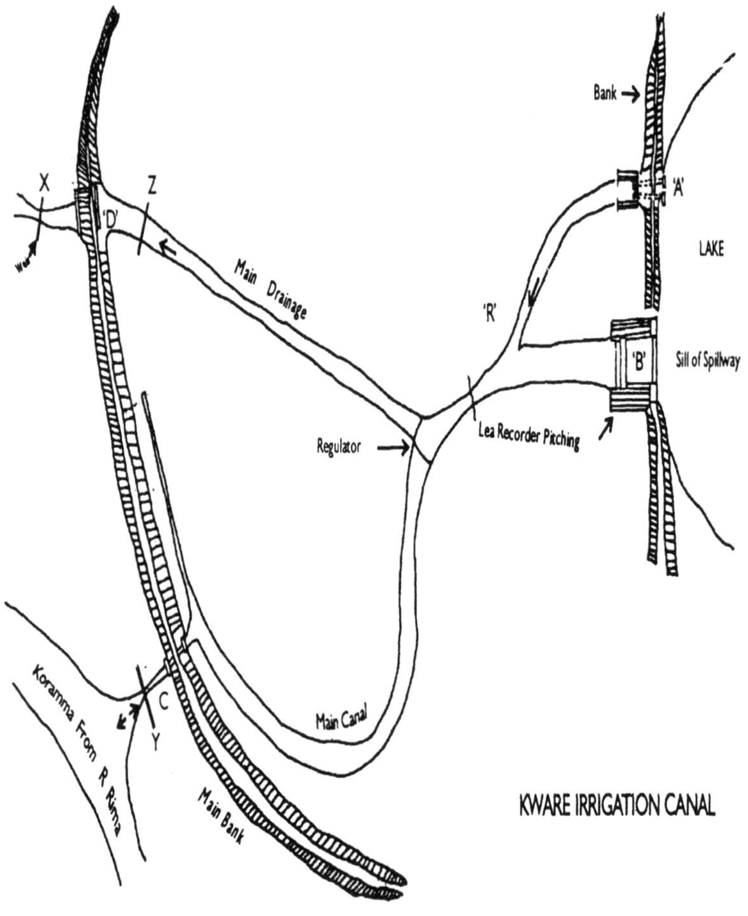

Fig. 4.2: A sketch of the canal, main drain and embankment system of Kware.

At "C" a 91cm (3') diameter culvert was used to let water flow into or out of the irrigated area. Either end of the culvert could be closed partly or fully by means of "timber needles" about 15 cm (6") wide. Flood water from the Rima river could be allowed into the irrigated area when it was necessary to inundate the entire area. "D" is the main drainage outlet made up of 3 – 91 cm (3') diameter culverts with "laps" fitted to the outfall end of the culverts. Water flowed through the pipes when the level of water inside the bank was higher than the level of water outside. The "laps" prevented water from flowing from outside the banks into the irrigated area. Figures 4.3 – 4.6 give the details of the structures at points A,B,C and D (KMAG, File No. 1060 Vol. II, 1932).

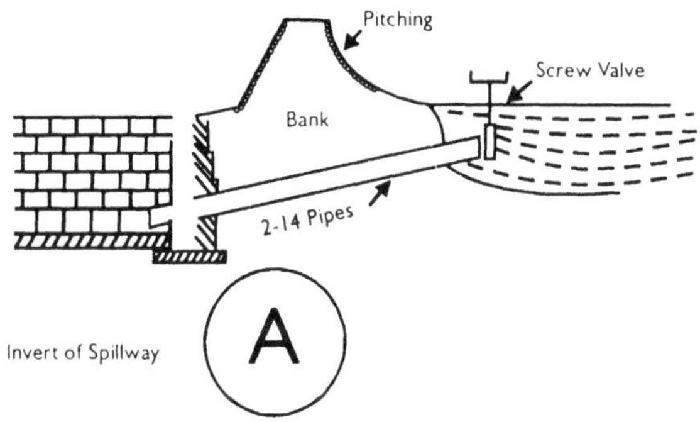

Fig. 4.3: Cross-section of the structure at A

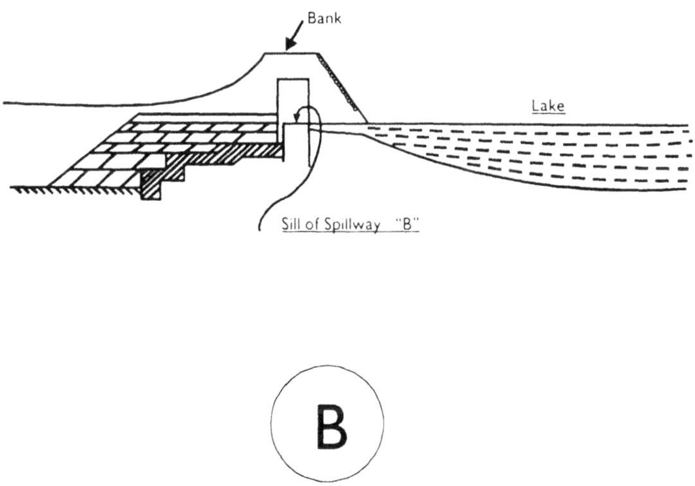

Fig. 4.4: Cross-section of the structure at B

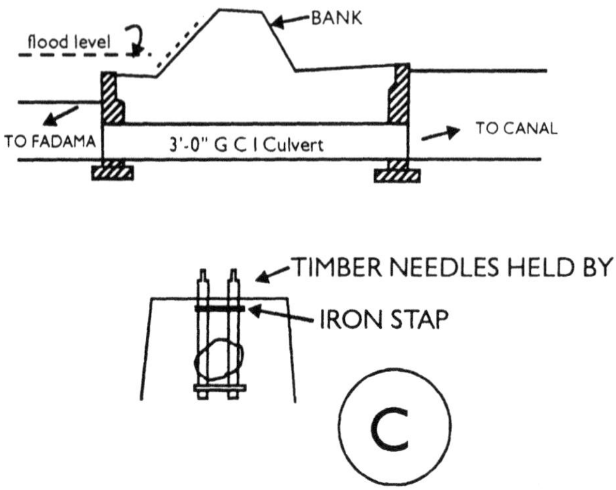

Fig. 4.5: Cross-section of the structure at C

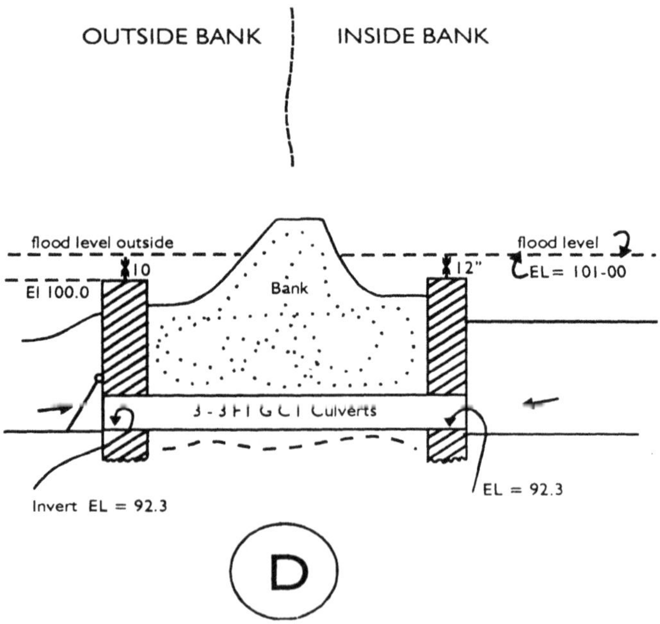

Fig. 4.6: Cross-section of the structure at D

4.2.3 Construction

The construction of the Kware Irrigation Project was to be progressive depending largely on demand. The work was initially supervised by Col. M.R. Collins, the first irrigation engineer to be engaged in Sokoto province.

The construction work started with the irrigation canal followed by the drainage canals; the construction of the Marabawa Flood Control Embankment started later. By June 1926, about 500 m of the irrigation canal had been completed. However, by March 1926, the head regulator and some length of the irrigation canal had been completed and the project was formally opened on this date by W. Ormsby-Gore, the Under–Secretary of State in the presence of Sarkin Musulmi, M.P. Rhys, the Resident, Sokoto province, M.R. Collins, and others. "Water is now flowing steadily throughout its length, and is available for irrigation as soon as the people care to use it, and excavate the necessary watercourses to their lands. There was great enthusiasm among the people at the passing of the water" according to Collins (KMAG, File No. 82/1924/56, 1926). Trees were planted to celebrate this event by the Under-Secretary of State, Sarkin Musulmi, and the Resident.

In May 1926, Mr. A.L.V. Long, an irrigation engineer, was appointed to replace Mr. Collins. He continued the work on the Marabawa Flood Embankment and completed it on the 3rd of July of the same year. He spent the dry season of 1928 to provide the drainage system for the scheme. The objective of the drainage works was to lower the water table in the area commanded and to reclaim the swampy areas. In November 1928, he started the inundation scheme and completed one third of it by the end of the year. After Mr. Long, there was a gap of more than 20 years before another irrigation engineer, Mr. J. Kitching, was appointed in Sokoto province in June 1949.

4.2.4 Agriculture

a) The Experimental Farm

Mr. Collins had proposed the establishment of a Government Agricultural Station as an important component of the Kware Irrigation Scheme. The Experimental Farm was to be used for experiments on irrigated agriculture and to obtain information and experience necessary for training the local farmers on irrigated agriculture.

Work on the Experimental Farm started in 1926 and by 1927, 17 ha of land was being irrigated at the farm. Between 1926 and 1933 the most important experiments in the farm were as follows:

- selection of improved varieties of rice, wheat and maize, among other crops;
- testing the effects of late planting and the effects of varying amounts of water at varying intervals;
- use of cattle and cattle drawn implements for cultivation and for land levelling;
- ploughing and levelling land under water; and
- puddling the soil and then transplanting of rice.

In 1934, the Native Administration took over the Experimental Farm from the Department of Agriculture, and while the rice selection trial continued, the investigation of the economics of rice growing by mixed farming methods became the dominant experiment for a number of years. As a result of these experiments a system of mixed farming in fadama land was introduced to farmers in the province (KMAG, File No. 1060 Vol. III, 1934).

b) The Project Farm

The land for the Kware Irrigation Project was developed by both the Native Administration (N.A.) and the farmers themselves. The N.A. undertook the development of the dry land as well as the reclamation of the swamp land. By 1927, about 120 ha of land was developed, 20 ha of this being developed by the farmers themselves. About 17 ha was allocated to the Experimental Farm and the remainder was cultivated by the farmers. The farmers whose lands were expropriated for the structures and for the Experimental Farm were settled on the reclaimed land. By 1934, about 93 ha of land had been expropriated and the owners resettled. In 1929, 300 - 400 ha of land was available for irrigation.

4.2.5 Management of the Scheme

Each crop farmer was allocated 0.40 ha of land while mixed farmers were given 1.2 to 1.6 ha each; mixed farmers were also advanced cattle and implements. The main dry season crops were wheat, cassava, sweet potato, tobacco and sorrel while rice was the main rainy season crop. The seasonal water application was about 0.7 l/s/ha for rice.

If a farmer could not cultivate all his land effectively every year, then the excess land was given to those who had no land. As far as possible, a new farmer was allotted a plot of land in his district. With the agreement of the district heads, vacant plots could be cultivated by farmers from other districts. The district heads supervised the distribution of seeds to farmers. The cost of these seeds were subsidized with funds from the Native Administration. Farmers who could not pay for the seeds and other inputs at the beginning of the season did so at harvest either in cash or in kind. The district and the village heads were consulted also in cases of boundary disputes. In 1956, there were 600 farmers in the scheme.

In 1934, as indicated above, the Native Administration took over the scheme particularly the Experimental Farm from the Department of Agriculture. The responsibility for the maintenance and operation of the scheme devolved on the Native Administration until 1956 when a Local Management Committee of 22 members was set up under the chairmanship of the district Head of Kware. The irrigation water rates of 5-, 8-, and 10 shillings were charged depending on the "amenities" offered to the various parts of the project (KMAG, File No. IRR 104, 1958).

Requests were made for the scheme to be declared a Settlement Scheme or the area to be declared a Settlement Area under Regulation 1 of 1950. This would provide a legal basis for the scheme and a legal power to charge economic rates for water and other services. This was particularly important to the Management Committee because it had no legal means of enforcing the payment of rates. Palmer-Jones (1981) reported that agreement was reached with the Sultan of Sokoto and his Council in 1959 to declare the area a settlement area but it was not until 1963 that Kware was declared a settlement area.

Kware Irrigation Scheme was considered to be a success during the first few years. At the end of the first year, valuable preliminary experience was gained. Cattle were used successfully for ploughing and for land levelling and the crop yields were considered to be "fair" Probably the most successful year of the scheme was 1931. The protective bank achieved its purpose; the water levels in the fields were controlled according to the plan; the yield of rice was high enough to cover the cost of production by hired labour, reaching the very high level of about five tonnes/ha (previously the best yields ranged from 2.5 - 4 tonnes/ha); the interest and confidence of the farmers in the scheme reached a very high level (Agricultural Department, Nigeria, 1931). However, the scheme saw more failures than successes; these are adequately covered in Section 4.3.

4.3 Problems of the Scheme

As indicated in section 4.2.3, the Marabawa Flood Embankment, which was expected to protect the scheme from the Rima floods, was completed on 3 July, 1926 but in 1927 and also in 1928 the embankment was breached due to flooding. In 1930 the embankment burst and all the crops in the Experimental Farm were lost and in 1933, Mr. J.H. Palmer, the Superintendent of Agriculture, had to breach the bank to release the surplus water. There were other breaches in 1936 and 1939. However, in 1940 the protected area was inundated not by the Rima floods but by internal floods originating from the Shalla lakes. At this time, the flood scheme was abandoned and the project continued to function only as a dry season irrigation scheme.

Seepage and deep percolation from the water supply canal were really very high. It contributed not only to the rise in water table in the project area but obviously it reduced the amount of water that could be used for irrigation. The seepage and deep percolation losses between the canal head and the irrigated field over a distance of 1.6 km were said to be 26 per cent at full supply level and 15 per cent at 15 cm below the full supply level (KMAG, File No. 30867, 1926).

From the point of view of the colonial administrators, another serious problem of the scheme was the poor participation of the farmers in it. Farmers were not taking up land in the scheme as expected and some of those who took up the land did not work as hard on them as could be expected. This is why some of the administrators decried the lack of legal backing to force the lazy farmers to cultivate their lands properly and to expropriate the land within the scheme. However, it has been strongly argued that "force" would not have improved the performance of the scheme (Palmer-Jones, 1981). It could have been surprising if the constant flooding of the scheme and the consequent loss of crops did not affect the attitude of some farmers. Besides, tilling and levelling the land with cattle, puddling the land and transplanting of rice required a lot of personal effort and hired labour. Also, labour itself was scarce and the market for the products of the scheme was not assured. Surely, the attitude of the farmers could be ascribed more to technical and economic reasons than to lethargy.

In January, 1944, the Government of Nigeria requested the Government of Sierra Leone to permit Mr. C.J. Rae, an irrigation and drainage engineer, to visit Nigeria and study the "Possible Agricultural Development by Means of Irrigation and Drainage in the Colony and

Protectorate of Nigeria." Following this invitation, Mr. Rae spent February and March, 1944 visiting nearly all the provinces of Nigeria to assess the irrigation situation in the country. Kware Irrigation Scheme was the most interesting project he visited. He described the scheme as a complete failure. This failure, he noted, had created a strong prejudice against any further irrigation work in Sokoto Sultanate. To him, it was extremely important to prove Kware scheme successful. He, therefore, recommended a thorough rehabilitation of the Kware project (Rae, 1944).

Rae attributed the failure of the scheme to two reasons:

- defective design, and
- no technical supervision of operation and maintenance of the scheme.

In terms of defective design, the flood embankment was too low and did not conform to the flood gradient of the Rima river. The alignment of the main canal and its grading were poor and so also were the arrangement by which the Shalla water was drawn off for irrigation and the internal irrigation water distribution system. In terms of supervision, it appears that the technical staff to supervise the operation and maintenance of the scheme were withdrawn soon after the completion of the original construction. This appears to be supported by the fact that in 1928 progress at the scheme was again hindered due to the ill-health of the officers stationed there and by the fact that after Mr Long, there was a gap of more than 20 years before another irrigation engineer was appointed in Sokoto province (KMAG, File No. 4637, 1953).

It appears that the flood scheme was resuscitated between 1940 and 1947 because in 1947, three years after Rae's visit, the Marabawa bank broke in seven places due to exceptionally high water and another suggestion was made to abandon the flood scheme. Probably because of Rae's suggestions 10 years earlier, the scheme was modified by the Irrigation Department in 1954. The embankment was regraded, the breaches repaired, and a spillway was provided for the Shalla floods. In 1955, the embankment was repaired again and all the irrigation and drainage channels were cleared by the Sokoto Native Administration. However, in 1958 the scheme had again deteriorated very badly; the irrigation and drainage channels were blocked; the land had become waterlogged; and maintenance had been neglected completely. Some feeble efforts may have been made to maintain the scheme between 1958 and 1963 but after 1963 no further efforts were made to resuscitate the scheme until the 1970s (Palmer-Jones, 1981).

4.4 Lessons from the Scheme

The Department of Agriculture which planned and executed the Kware Irrigation Scheme learnt a lot of lessons from the scheme. As a result of this scheme, the Department was in a position to give advice on several aspects of irrigation in Northern Nigeria. Specifically, the department could advise on (KMAG, File No. 1060, Vol. III, 1934):

- the best means of irrigation and drainage water control,
- the best method of growing rice (use of bullocks to prepare and level land; puddling the soil; and transplanting rice),
- the best crops to grow during the dry season,
- land suitability for crop production,
- cropping intensities, and
- maintenance of soil fertility, among others.

The above lessons were important and relevant to the Department of Agriculture at that time, but of more relevance and importance are the lessons from Kware for the present day planners, developers, managers and operators of irrigation and drainage systems in Nigeria. Some of the lessons from Kware may be summarized as follows:

i) Sound design (both conceptual and engineering) and careful construction are essential for a successful scheme. Defects in design contributed greatly to the failure of Kware. Examples still exist of modern projects with these problems.

ii) The construction of the Kware Irrigation Project was progressive depending largely on demand. Some lesson could be learned from this concept. There is no reason to develop say 40,000 ha of land at once when the capacity to crop the land is at the incremental rate of say 2,000 ha per annum,

iii) A research farm, established at the same time with the project farm, was an important component of Kware. Irrigation research and extension have not yet been given serious attention in Nigeria.

iv) No serious attention was paid to the operation and maintenance of Kware. "This is what has killed the Kware Scheme" (Ray, 1944). Lack of serious attention to operation and maintenance is still "killing" irrigation schemes today.

v) Farmer participation can make a better contribution to a successful irrigation project than farmer control. A lot of effort was made in trying to "force" the "lazy" and "uncooperative" farmers to perform in Kware. These efforts were not very successful.

vi) Security of land tenure is important to the farmers in any irrigation project. The farmers at Kware were always under the threat of expropriation of their land for poor performance or for uncooperation. Nwa (1994) recommended that land ownership by the farmers, must be recognized as one of the essential conditions for the success of Water Users Association in irrigation schemes.

vii) At the project management level, farmers must be given not only duties but also rights as well. Kware farmers were expected to pay tax, pay for water and other inputs, and to cultivate their land by "force" if necessary, but if they lost all their crops through the breach of the Marabawa embankment, that was "too bad." Nowadays, if a farmer loses all his crops due to late planting as a result of late land preparation (for which he had paid) by the project management, that also is "too bad" for the farmer.

viii) There is need to understand the real economic situation of the small farmers in an irrigation project. At Kware, the intensive work of land preparation and levelling with cattle, puddling and transplanting of rice, shortage of labour, and the problems of marketing, may have forced the farmers to take the economic decision not to participate or cooperate in the scheme. From his field studies, Palmer-Jones (1981) concluded that "provided there is a real economic incentive for these farmers to undertake the actions required by the scheme and provided there are means of risk insurance and conflict arbitration, small farmers will willingly participate in irrigation schemes."

Chapter 5

Irrigation Development Other than Kware (1925 – 1959)

5.1 General

The period 1925 – 1959 witnessed two great events in the world, the effects of which did not bypass Nigeria. These were the great depression (1929 - 1933) and the Second World War (1939 - 1945). Just as Nigeria suffered some adverse effects from the above events, she also benefited from the post-war boom which followed. During the boom period (1945 – 1960), groundnut production was expanded and pioneer oil mills and marketing boards were introduced.

Also during this period, the ten-year development plan (1946 - 1956) was launched for Nigeria. The above events and activities had their effects to varying degrees on irrigation developments discussed in this chapter.

5.2 Survey of Conditions for Irrigation in Nigeria by C.J. Rae, 1944

Mr. C.J. Rae was an irrigation and drainage engineer based in Sierra Leone. In 1944, the Government of Nigeria requested the Government of Sierra Leone to permit Mr. Rae to visit Nigeria to study the conditions for irrigation in the country. Following this invitation, Mr. Rae spent February and March of 1944 visiting nearly all the provinces of Nigeria to assess the irrigation situation.

In Niger province, he visited Bida, Wuya, Edozhigi, Badeggi, Jima, Agaie and Abuja (now Suleija). He was impressed with the large fadama land along the banks of the Niger, Kaduna and Bako rivers and with the "keenness" and "water-mindness" of the local people who had already been carrying out a great deal of irrigation entirely on their own initiative. At Edozhigi, a small irrigation scheme was already in use and at Jima, the local people had dammed the main stream at several points, taking off irrigation channels, some several kilometres long, to irrigate very large areas, 2,000 to 2,500 ha. He estimated that over 81,000 ha of fadama land could be found in the Niger-Kaduna-Bako flood plains to be developed for irrigation particularly for irrigated rice.

He specifically recommended the Wuya-Edozhigi and Badeggi areas as good locations to start any such development (Rae, 1944).

In Zaria and Katsina provinces, he noted that a large number of places around Shikka near Zaria and the extensive fadama around Maigana would be particularly suitable for "tank irrigation." He observed the small land settlement scheme in Daudawa which he thought could be used to study the question of the settlement of reclaimed or improved lands.

Rae took most of his time in Sokoto province but he still managed to visit Argungu, Birnin Kebbi, Jega, Gusau, Gwandu and Sokoto. He noted the fadama land in those areas and took a particular note of the Gwandu fadama which was very intensively cultivated mostly with sugar cane.

In Kano province, he visited the region north and west of Kano including Minjibir, Danbatta, Bichi, and Dawaki-ta-Yanma districts; east of Kano including Gezawa, Gabasawa and Ringim districts; south of Kano as far as Tamburawa; and Kano City itself. He saw the oxbows of Jakaradi river, the intensively cultivated areas along the Challawa river, and the extensive vegetable gardens on the northern fringe of Kano City. Since these vegetable gardens were close to the city, he suggested that electrically driven pumps could be tried to replace the shaduf; he also suggested the trial of animal driven water lift system which would still be less tedious than the shaduf used by the farmers.

Rae's visit to the Southern provinces took him to Afikpo, Unwana and Umuahia in the present Ebonyi and Abia States; Umon and Itu in Akwa Ibom State; Onitsha in Anambra State; Ubiaja and Illushi in Edo State; Owena and Okitipupa in Ondo State; and Ibadan in Oyo State. In his canoe ride down the Cross River from Afikpo to Itu, he saw the fertile valley swamps similar to the inland swamps in Sierra Leone which could be reclaimed for rice cultivation. He also saw similar areas in the Owena and Okitipupa swamps, and at Illushi on the right bank of River Niger, he estimated that about 11,000 ha of the flood plain could be developed for rice cultivation.

In summarizing his visit, Rae noted the tremendous potential in Nigeria where irrigation, drainage and land reclamation works could be undertaken to boost the food production of the country. However, he observed that apart from rainfall data, there was a serious lack of topographic maps and hydraulic data (flood levels in rivers, maximum and minimum discharge records of rivers, runoff data from rural watersheds, etc.) upon which the design of such works could be based.

In his recommendations, he stated that data collection should receive immediate attention. He made specific recommendations that development work should start in the following locational order of priority:

1. Bida district of Niger province,
2. River flood plains along the Niger in Ilorin province, and Illushi (Benin province),
3. Sokoto province,
4. Minor development in Kano province, and inland valley swamps in the Southern provinces,
5. Cross River basin, and
6. Littoral swamp areas.

He gave three principal considerations on which he based the above priority. These were:

1. The readiness of the local farmers to take advantage of any facilities granted by the works,
2. The availability of manpower for the settlement of reclaimed and developed areas, and
3. The existence of transport facilities for the movement of produce to the marketing centres.

5.3 Rice Mission and the Development of Rice Production in Nigeria

The central Niger Delta has been mentioned as the original centre of variation of the indigenous West African rice species (the red rice), *Oryza glaberrima steud.* The spread of this rice into the fadama lands of Northern Nigeria during the 16th century is said to have initiated rice cultivation in Nigeria (Hardcastle, 1959). Although the evidence to show who introduced white rice, *Oryza sativa L.* into Nigeria has not been found, the upland varieties of the crop were introduced into the high forest zone of Western Nigeria after about 1879. However, the Department of Agriculture made the earliest recorded introduction of the white rice into Nigeria after 1919. The department introduced the shallow swamp varieties from British Guiana and Ceylon into the fadama lands along Kaduna, Niger and Benue rivers and their distributaries where they rapidly replaced the red rice (Hardcastle, 1959).

In 1933, Mr. R.R. Glanville of the Sierra Leone Department of Agriculture, was asked to visit Southern Nigeria to study the possibility of growing rice in the mangrove swamps as it was the case in Sierra

Leone. Mr. Glanville carried out his first experiment in Warri. In 1934, he planted a small plot near the rest house in Warri and obtained a paddy yield of about 3,000 kg/ha, and in 1935, 11 local farmers started growing rice in Warri (Agriculture Department, Nigeria, 1935); in 1936 a total of 22 farmers were growing rice in the Jekri - Sobo District of the Warri Division. Experiments carried out on tidal mangrove swamps near Calabar and Oron gave satisfactory results but the local farmers were reluctant to adopt rice farming (Agriculture Department, Nigeria, 1936).

Rice was first introduced into Ogoja province, i.e., Abakaliki, Afikpo and Ogoja divisions and Obudu sub-division in 1943 and good progress was made in expanding the area planted in Ogoja division and especially in Abakaliki Division. In 1945, demonstration and trial plots of swamp rice were laid down in Okigwe, Aba and Ahoada divisions while trial plots were established in the inland swamps in Abak, Ikot Ekpene, Arochukwu, Ndiya and Eket. In Onitsha province, plots were established in Ogurugu, Onitsha, Awgu, Awka, Udi and Nsukka. In Aba division, most of the swamps were extensively used for rafia palms; the people showed little interest in rice cultivation; also Awgu people did not show much interest. The results were poor in Ikot Ekpene but good yields were obtained at Arochukwu while Okigwe gave the best results and farmers also responded with enthusiasm (National Archives, 1945).

Rice was grown largely in the fadama areas of the Northern provinces and in the inland valleys in the Southern provinces. The estimated level of production by province in 1944 is given in Table 5.1 (Agriculture Department, Nigeria, 1945).

One of the major rice growing areas in the country was Lafiagi division in Ilorin province. In 1947, there was a discussion to extend the rice production area in the division by building the Egwa river dam. Unfortunately, the project was stopped by "lack of aerial photographs and funds" (KMAG, File No. 42627, 1947). Figure 5.1 shows the rice areas in Lafiagi division and the proposed dam site.

Through the efforts of the various provinces, rice production throughout the country expanded fairly rapidly so that by 1950/1951, rice had become the sixth major domestic food crop in Nigeria behind cassava, yam, guinea corn, millet and maize, in that order when approximately 200,000 ha were cultivated (Hardcastle, 1959).

Table 5.1: Estimated level of rice production by province in 1944

Province	**Rice production, tons of paddy**
Sokoto	21,690
Benin and Warri	12,000
Niger	4,850
Ilorin	3,630
Kano	3,000
Onitsha	2,500
Abeokuta	2,500
Adamawa	2,300
Kabba	1,470
Ondo	1,100
Zaria	900
Bauchi, Borno	810
Ogoja, Oyo, Colony, Owerri	510
Total	56,260

To support the rice production efforts in West Africa, the West African Rice Research Station was established at Rokupr in 1945 in Sierra Leone to serve all the British West African Territories. But by this time, Nigeria's rice production had increased tremendously and there was a felt need to establish rice research stations within Nigeria. Consequently, the Federal Rice Research Station was established in 1953 at Badeggi in Niger province. The work of the station was concentrated on yield and crop quality improvement, seed multiplication, soil fertility and cultural practices investigation all in close liaison with the West African Rice Research Station in Sierra Leone (Hardcastle, 1959).

5.4 Small-Scale and Village Irrigation Schemes

In 1937, the Colonial Governor of Nigeria in Lagos felt that "the time will sooner or later arrive when serious consideration will have to be given to the possibility of irrigation in the Northern provinces" and, therefore, there was the need to collect information ready for such an eventuality. Consequently, a sort of survey of the situation of irrigation in the Northern provinces was ordered by the Secretary, Northern provinces, when he wrote to the Residents of all the provinces. His letter stated in part:

> I am to ask you to state what areas you consider suitable for irrigation by means of river storage, whether records of river levels exist in those areas and, if not, what measures you propose should be taken for the keeping of such records in future. I am to ask also (to Kano and Sokoto residents only) for a brief summary of the information already available regarding the possibilities of irrigation from the Sokoto and Hadejia rivers (KMAG, File No. 26902, 1937).

Of all the 12 Northern provinces, it was only Adamawa province that could not give an immediate response because the Resident needed time to discuss the matter with the Deputy Director of Public Works. Various responses were received from the other provinces; for example, there were no areas in Bauchi province where large scale irrigation could justify the cost. Benue did not need irrigation. In Borno province irrigation farming was already important in the following areas:

i) the Komadugu Yobe (from Gashegar to Lake Chad),
ii) the western shore of Lake Chad, and
iii) certain parts of the course of the River Gongola.

And in Ilorin province irrigation was practised on a small scale in the following areas:

i) the Niger marshes between Jebba and Pategi,
ii) between Bussa and Rofia in the Borgu division, and
iii) in the marshes of the Asa River close to Ilorin.

"But it is doubtful whether the local peasantry would take much interest in a large irrigation scheme."

Kabba, Katsina and Zaria provinces suggested certain locations and rivers where irrigation could be undertaken. Irrigation was already important in certain parts of Kano but additional areas were all the same suggested. In Niger province, irrigation was already practised on a small scale in the open valley near Ndeji Gwazan (Lemu district); Agaie and Badeggi flats and the Zugurma area of Kontagora were suggested as additional areas. In Plateau province, tobacco was the only dry season crop grown by the local population at the time, and it was irrigated by hand. Riyom river and several other areas were suggested as being suitable for damming. Sokoto province reported on Kware

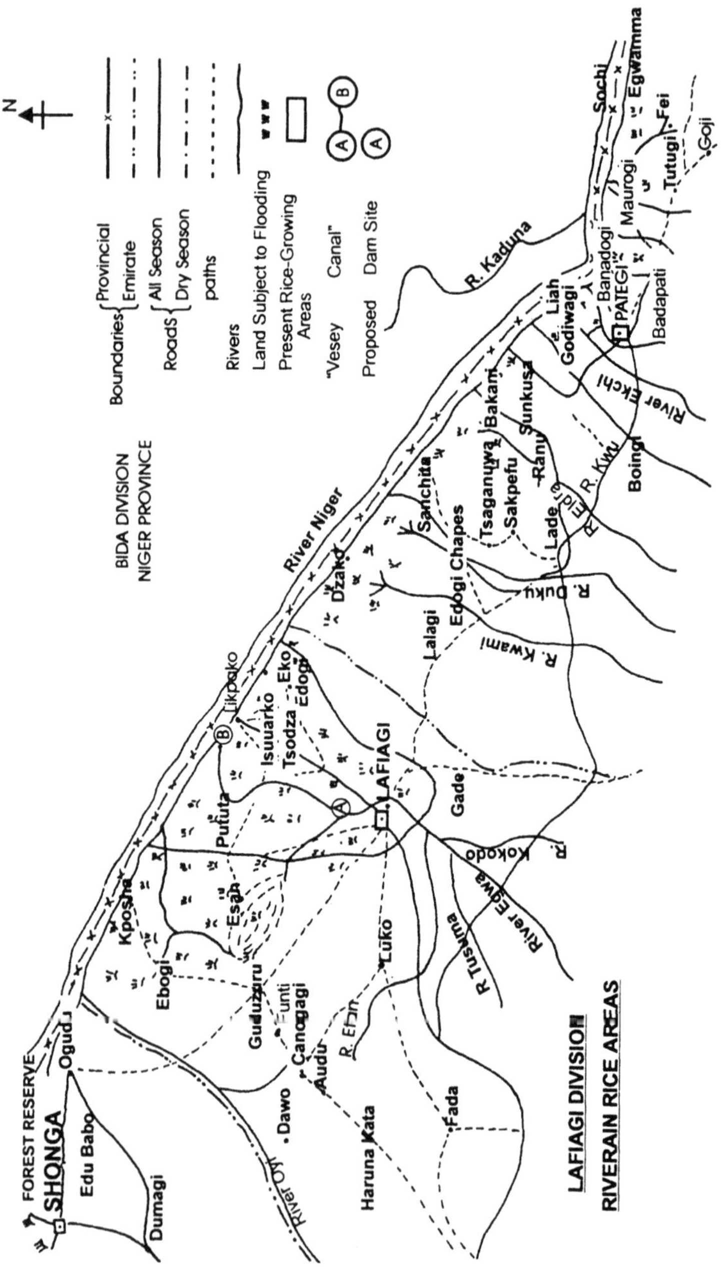

Fig. 5.1: Rice areas in Lafiagi division and Egwa dam site

and recommended canal irrigation on the Jega area of the Zamfara river but then emphasized that "there should first be an agricultural report and then a detailed survey."

Between 1926 and 1959 a number of small nurseries and orchards were established under irrigation in the Northern provinces. The Kazaure nursery which was 0.8 ha and planted to mangoes and citrus, and was irrigated by hand from wells, was opened in 1926. The Bauchi nursery and orchard, installed in 1957 was irrigated by the shaduf. The Katagum nursery and orchard were established the same year at Shira. Two nurseries and orchards were irrigated from storage tanks supplied by 18 m windmills. One was the 2.2 ha Jakara nursery and orchard (near Kano) which was established in 1953 and planted to oranges, grapes, tangerine and tangelo, and the other was the 0.4 ha at Danbatta also planted to citrus (Kano Report, 1957).

In 1943, local farmers constructed a weir across Riyom river in Plateau province and built the Riyom farm for potato production during the dry season. In that year 20 farmers operated the system and in 1944 the number of farmers had increased to 70 due to the profitability of the system. Also at Ta Hoss near Riyom about 60 local farmers established an irrigation scheme in 1943 under the guidance of the agricultural officer. The scheme was irrigated by gravity with water from springs. A pilot irrigation scheme of 16 ha was later in 1959 initiated at Ta Hoss.

Also in 1943, small farms to irrigate Irish potatoes were established in Toro and Bauchi. During the same period, minor irrigation schemes were started at Dala Waja, Shabbal and Kanawa in Gombe Emirate where local farmers were taught how to run irrigation farms using the shaduf, and at Darazo in Bauchi division for training ex-servicemen in irrigation farming (Agriculture Department, 1944; Agriculture Station Report, 1947).

In 1952, 6 village irrigation schemes were constructed in Sokoto province as follows:

1. Gwotomo, near Argungu, 30 ha; water for irrigation came from perennial streams;
2. Gummi, 20 ha; storage dam and drainage project;
3. Chakkai, near Kainji cattle dam project;
4. Dangwarema, near Kalgo, 40 ha; swamp drainage project;
5. Gawassau, near Argungu, 40 ha; swamp drainage project; and
6. Farfajia, near Jega, 280 ha; flood protection project.

Construction was to start on the following:

1. Shinkafe; 57 ha; storage dam and drainage;
2. Geben Kabasha, near Chafe; cattle dam;
3. Gusau farm; cattle dam; and
4. Bukwium; 57 ha; storage dam and drainage.

(The construction of Bukwium actually started in 1953).
And investigations were to start on the following schemes:

1. Wamako (Sokoto Emirate),
2. Gaminda valley (Sokoto Emirate),
3. Upper Dangwarema (Gwandu Emirate),
4. Zamfara valley tapki, and
5. Kefi, near Gada (KMAG, File No. 4637, 1953).

Other village irrigation schemes were Loguma and Toroko. Loguma irrigation scheme in Agaie in Bida division which was actually proposed in 1949 was to be 125 ha. Construction started in 1956 and was completed in 1957. The first irrigation also took place in 1956. Water was diverted from Emi River to irrigate rice which was the main crop of the project. The construction of Toroko scheme near Toroko in Niger province started in 1953. An area of about 80 ha was proposed for the project which was to divert water from Kanko river to irrigate rice. First irrigation took place in 1954.

5.5 Water Pumping from Wells and Small Streams for Irrigation

Between 1926 and 1959 a number of small nurseries and orchards were established under irrigation in the Northern provinces. The Kazaure nursery which was established in 1926 was irrigated by hand from wells. The Bauchi nursery and orchard established in 1957, used shadufs to pump water from wells. In 1946, a cattle operated water lift was used experimentally in Bauchi while a persian wheel imported from Sudan to Bauchi also in 1946 was able to irrigate 2 - 2.5 ha from a depth of 10.7 m The Department of Agriculture in 1932, introduced bullocks for lifting water from wells and small streams for irrigation into Enugu Zone and other parts of Nigeria. The farmers were very appreciative of the method (Agriculture Department, Nigeria, 1933). In 1975, a 2 - bucket water lifting device, powered by a single ox and using pulleys, was constructed and tested in Agubu near Hadejia. This device called " Super Shaduf" was able to lift water from four metres to irrigate vegetable crops.

In order to test the possibility of using wind mills for pumping water in Northern Nigeria, a recording electrical anemometer was installed in Sokoto near the Native Administration offices in 1931. In 1932, a small windmill pumping machine was installed near the Katsina Native Administration building while a second mill was installed on a well at Azare hospital in 1933. The outputs of these mills were compared to the wind speeds at the Sokoto station. It was found that a minimum wind speed of 8 km/hr was needed to operate the wind mills and that the wind speed was less than this rate in about three weeks per month (Pedder, 1937). By analyzing wind speeds for pumping water in Zaria in 1981, Nwa (1981c) found that the wind speed was less than 8 km/hr for an average of three weeks per month during a dry year and 1.4 weeks per month during a wet year. The conclusion by Pedder (1937) and Nwa (1981c) was that a storage for about three weeks supply of water would be needed for a dependable water supply for irrigation or domestic use from wind machines particularly during dry years.

Water pumping from wells and small streams for irrigation did not become fully established in Nigeria until the advent of the National Fadama Development Project (NFDP). This is discussed in detail in Section 6.6.

5.6 Construction of Small Earth Dams, Tanks and Tapkis for Irrigation

A large number of medium and small size dams, tanks and tapkis were built in Nigeria between 1925 and 1959. The majority of them were for cattle watering and for human consumption. A few of these include Kogin Giri in Jos, constructed in 1935; Okene dam, 1937; Eleyele in Oyo State, 1942; Bosso in Minna, 1946; Tenti in Plateau State and Agba in Ilorin, 1949; and Lamingo in Jos, constructed in 1957. An interesting example of a tank which was used for both village water supply and for cattle watering was the one in the Boho valley near Wababi in the Dange - Wababi area of Sokoto province. This tank was cleaned out and deepened by communal effort under the direction of Sarkin Baura in 1942. The tank was first constructed by Sarkin Rafi, the District Head of the area in 1900. According to local belief, it was near the site of the tank that Bello, the Sarkin Musulmi, found water when his army encamped at Wababi before 1900. In 1950, the tank was reconstructed and could hold about 460 m^3 of water. The tank was supplied with water by surface flow during the rainy season and by natural springs in the floor of the tank during the dry season (SOKPROF, File No. 6613, 1950).

An example of a small irrigation dam constructed during the period was the Gummi dam in Sokoto province which was constructed in 1952. The dam was 2.4 m high and 800 m long and provided water to irrigate 20 ha of land. The Bukwium irrigation dam also in Sokoto province was constructed in 1953 (KMAG, File No. 2708, 1951). The Egwa river dam was proposed in 1947 for rice irrigation in Lafiagi division but was not constructed.

5.7 Medium and Large-Scale Irrigation Projects

5.7.1 Investigations and Surveys for Irrigation Projects

Investigations for irrigation projects in the Niger province started in the middle 1940s. The proposal for the Wuya-Edozhigi Irrigation Scheme was made in 1944 in Rae's report (Rae, 1944). The scheme was originally to be used as a rice research station but the idea was later abandoned. The investigation of the scheme started in 1946. Stream flows were measured daily for four years at Wuya and this information was used to design the scheme. In 1948 the area of the project was contoured and other design data were collected before the actual proposal of the project was submitted in 1949. For the Badeggi project, the survey work and the collection of the hydrological data started in 1949 in an area of the plain of Gbako river near Badeggi (KMAG, File No. 47678, 1949; and No. 4637, 1953).

In the neighbouring province of Ilorin, a district officer with an engineering qualification was posted in 1947 to Lafiagi to carry out preliminary investigations into the possibility of introducing irrigation schemes in the area for rice and other crops. In spite of the numerous small streams at Share, Shonga, Ogudu, Lafiagi, Lade, Pategi and elsewhere in the Nupe areas, action on any irrigation scheme was delayed because there was no irrigation engineer whose advice could be sought. In 1954, an irrigation engineer from Bida visited Pategi and selected gauging sites on all the streams being used for irrigation on the Pategi fadama. The gauges were installed and a gauge reader was sent from Bida to read them. A land survey team was sent also from Bida, to survey the area (Agriculture Station Reports, 1960).

In 1950, Mr. John Kitching, an irrigation engineer from Sokoto, visited Borno province and recommended highly the establishment of an irrigation scheme on the shores of Lake Chad. The Wulgo pilot irrigation scheme on the lake shore, was recommended. Two water level gauges were set up at Wulgo in 1950 and lake levels were taken weekly. Water samples taken at the location of the two gauges were sent to Lagos for analysis. The amount of dissolved solids were "minute" and ammonia and iron contents were "very high." A preliminary survey of the scheme was carried out also in 1950 (Agriculture Station Reports, 1960).

In addition to Borno, Mr. Kitching's pre-reconnaissance surveys of 1950 took him to Kano, Kebbi and Benue provinces. In Kano, he reported on the possible schemes along the Hadejia river and at Kazaure and Ringim. In 1953, an irrigation engineer, Mr. F.S. England, was posted to Kano; he continued the survey of the irrigation possibilities in the province.

In 1951, Mr. Kitching visited Calabar province from Sokoto. After a preliminary reconnaissance survey of two possible flood control cum rice irrigation project sites, the rice projects at Kwa Falls Estate, Calabar, and Lower Enyong Creek at Itu, he recommended that three flood gauges be established on the river at Kwa Falls Estate and that 800 ha of the flood plain should be contour-surveyed. At the Lower Enyong swamp, he suggested that 4,000 ha of land could be drained, protected from the Cross river and Enyong Creek floods and put to irrigated rice. He even submitted a proposed layout of the project (Fig. 5.2) (KMAG, File No. 2700 A, 1951; No. 4637/S.I, 1953).

In 1953, there was an investigation by the Kano Native Authority on the possibility of building a dam at the Tiga rapids on the Kogin Kano, near Rurum in Rano district. There was the need to clarify the purpose of the dam. Suggestions were made that:

1. the dam could provide valuable water conservation; the water could be released at the height of dry season into Kogin Kano to extend the flow of the river and improve the amount of water available to the Kano water works; this would augment the Kano urban water supply;
2. the dam could be used to produce hydroelectric power;
3. the water could be used for a fish farm; and
4. the water could be used for irrigation.

The final decision on the purpose of the dam was to await the production of the contour map of the area and the arrival of an irrigation expert to offer his opinion (KMAG, File No. 72390, 1953). Today, it appears that the dam is being used for all the four purposes even though irrigation is its most important function.

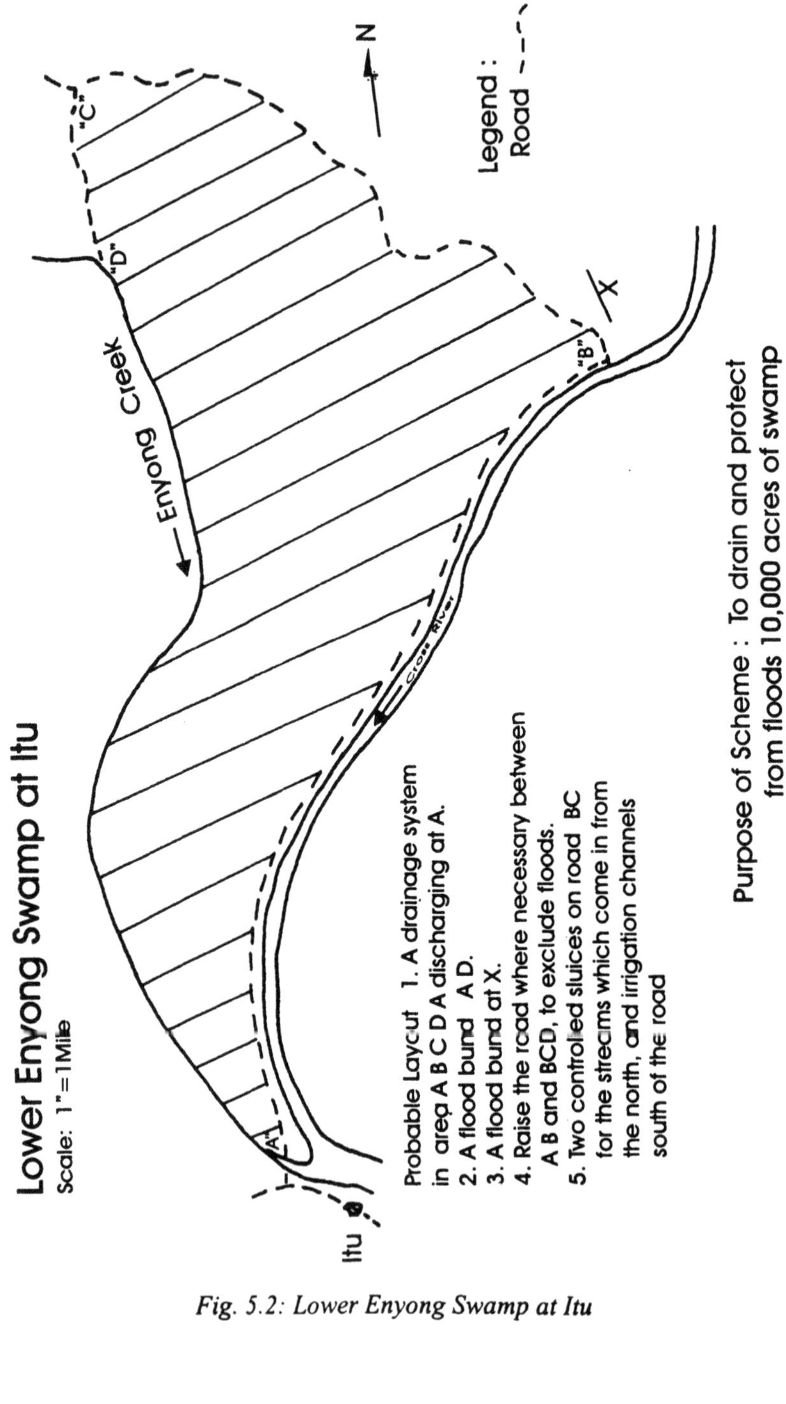

Fig. 5.2: Lower Enyong Swamp at Itu

A number of other surveys into the possibilities of irrigation in Kano were undertaken in 1956. These included:

a) Danbatta District:

i) dry season irrigation and flood control scheme on River Tomas,

ii) swamp drainage on River Gombe, a tributary of the Tomas, and

iii) irrigation scheme on Makoda lake.

b) Rano District:

i) Yuri dam

c) Dawakin Tofa District:

i) irrigation scheme on River Hogo

d) Other Districts

i) several earth dams.

In 1957, several stream gauging stations were established and surveys carried out along the Hadejia river from Kano to Wudil and a flood control scheme to protect about 600 ha of fadama land near Wudil was surveyed. A social survey revealed that a flood control scheme to make land available for arable farming would not be welcomed in Wudil, a predominantly Fulani community. The local people were afraid that they could lose their grazing rights. The three nearby villages opposed the Tomas scheme for the same reason. A number of fadama lands along the Challawa river from Kano to Ringim were also surveyed in 1958 (Agriculture Station Reports, 1960).

In Borno province the possibility of starting an irrigation scheme at Lake Alau was investigated in 1953 while the full investigational survey of the scheme was carried out in 1954. There were lots of activities in 1959 to try to start irrigation on a large scale in Borno province. Three areas were intensely investigated:

1. Marte (Firkin) area: The area was surveyed by irrigation surveyors and soil scientists. The soil chemist who analysed both the soils and the water of Lake Chad did not recommend the area due to the "certainty of salt accumulation."
2. Jillam area: The survey of soil and water were made and they showed that the area could be irrigated by gravity from the Ebeji river.
3. Yau area: After soil and water surveys, this area was said to have the best potential of the 3 areas and was given the

highest priority in spite of the fact that salt accumulation was considered to be a threat. However, it was reckoned that the river water could be used for leaching. Plans were made to start irrigation agronomy trials to deal with the salt problem and to investigate wheat varieties, rotation and cultural practices (Agriculture Station Reports, 1960).

5.7.2 Wuya-Edozhigi Irrigation Scheme

Wuya and Edozhigi are two small villages in the Kaduna river area near Bida in the Niger province. The Wuya-Edozhigi irrigation scheme was named after the two villages. The scheme was proposed in 1944 and the actual proposal with some preliminary designs was submitted in 1949; the proposed size was 1000 ha. The water for the rice scheme comes from two perennial streams, Kupanko and Edjiko; the two streams flow into Kaduna river. Kupanko flows through the middle of the fadama and Edjiko flows through one end of the project area. A flood protection embankment, protects the area from the Kaduna river floods. The embankment with side slopes of $1^{1/}{}_{2}$:1, and top width of 3 m, varies from 120 to 180 m from the Kaduna river.

The design called for irrigation outlets at suitable intervals in the bank of the canals; each outlet was to supply water for 40 ha. The irrigation network was designed to distribute water to one corner of each 10 ha rectangle. It was presumed that each farmer would be able to carry his water for 320 m without technical assistance.

Construction work was first started at the Edjiko stream end in 1951 while work on the Kaduna embankment started in June of the same year. "On the morning of 7th June, 1951, the King's birthday, the first main irrigation channel of the Wuya-Edozhigi irrigation scheme was opened by Muhammadu Ndayako, C.B.E., Emir of Bida". About 240 ha of land was available for irrigation at the time (KMAG, File No. 47678, 1951). By 1980, 850 ha of land was under irrigation at the scheme (Nwa, 1980). The layout of the scheme as proposed in 1949 is given in Figure 5.3.

5.7.3 Badeggi Rice Irrigation Scheme

The Badeggi Rice Irrigation Scheme was located in the flood plains of Gbako river near Badeggi. Unlike the Wuya-Edozhigi, the Badeggi scheme did not require a flood protection embankment, and, therefore, it was called a stream control type of scheme. When the proposed

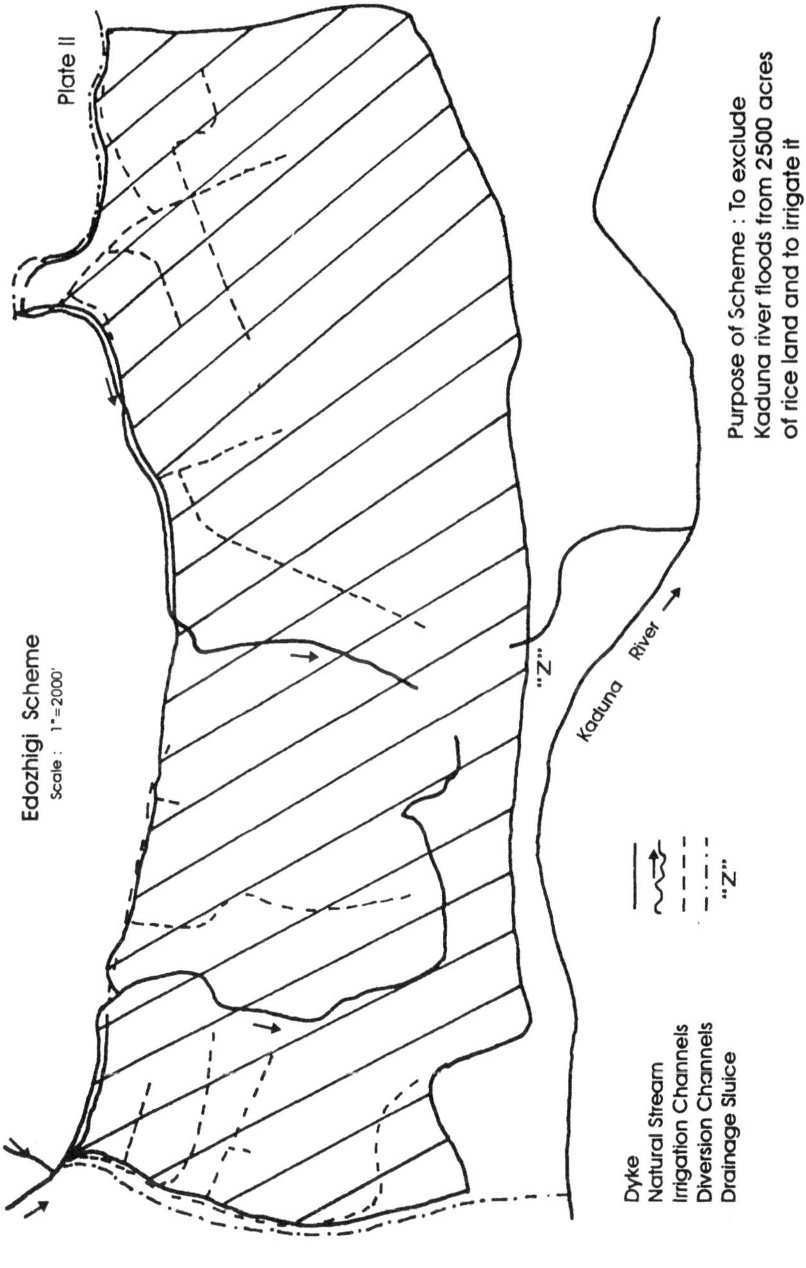

Fig. 5.3: Edozhigi Scheme

scheme was approved in 1949, it was expected that the scheme would irrigate about 1000 ha out of a gross area of about 1600 ha. Irrigation water is supplied from the head works on the Musa river and the fields are irrigated by gravity. The construction of the scheme started in 1951 and by 1955, about 650 ha of land was available for irrigation while 250 ha was actually irrigated.

The year 1955 appeared to have been a very important year in the life of the scheme. A Management Committee was set up for the scheme in 1955. The following were members of the Committee:

1. Col. A.A. Smith - Irrigation Engineer
2. Mallam Chado - Sarkin Gona
3. Mallam Tauhidi - Etsu Tafien
4. Mr. N.F. Cooke - Senior District Officer, Bida

In attendance: Mr. D.P. Drew - Production Officer.

The first decision of the Committee appeared to have been to declare the scheme a settlement area under the settlement regulation of 1950. This, among other things, gave the Committee the necessary power to charge water rate. In the same year the Committee appointed Mallam Abdullahi Alkali Agaie, as the first resident manager of the scheme.

There were about 500 farmers in the scheme in 1955. The farmers were asked to form a committee of "farmers' representatives" which would enable them to tackle their problems better, such as the problem of lack of water in some farm areas and the problem of pests which were rather serious in 1955. These must have contributed to the very low yields of paddy obtained by farmers that year i.e., 1230 to 1950 kg/ha with an average of 1600 kg/ha (KMAG, File No. 2705C, 1955). By 1980, 840 ha of land was irrigable in the scheme and 690 ha was actually irrigated (Nwa, 1980).

5.7.4 Duku-Lade Irrigation Scheme

Duku is a seasonal stream that flows into River Niger; the stream is about 0.8 km away from Lade village on the main road through the area. Lade is in Edu Local Government Area of the present Kwara State. Duku-Lade is a rice irrigation scheme which was started by the Irrigation Division of the Ministry of Agriculture of the Colonial Northern Nigeria Government. The survey for the project started in 1954 but the construction started in 1956. The scheme consists of controlling the flow of the Duku stream with concrete head works to divert the water for gravity irrigation of rice.

The proposed area of the scheme was 1000 ha and in 1959 when the head works were completed, about 80 ha of land was ready for irrigation (Agriculture Station Reports, 1965). By 1987, 800 ha was equipped for irrigation but only 150 ha was irrigated because the head works were in disrepair and the canals were not maintained (Ejieji and Nwa, 1989).

5.7.5 Wurno Irrigation Scheme

The main aim of the Wurno Irrigation Scheme was to protect about 1,500 ha of fadama land from the River Rima floods. The main structure of the scheme was an earth dam with a spillway 100 m wide and two outlets for two canals. The reservoir is filled from the flood spills of River Rima through a spill channel, the Balla creek. There are two irrigation main canals, the left bank canal (Lugu canal), and the right bank canal (Tududawa canal) to take water to the various parts of the field to be irrigated. The reservoir water was to be used during the wet season to irrigate the main crop, rice. The dry season crops were wheat and onion, among others.

The construction of the scheme started in December, 1957. In 1960, severe floods resulted in two large breaches in the dam; these breaches were later repaired. The construction of the scheme was completed in 1964. By 1980 the irrigable area was 800 ha and the area actually irrigated was 600 ha (KMAG, File No. IRR 104, 1959; Nwa, 1980).

5.8 Irrigation Policy for Northern Nigeria, 1953

5.8.1 Introduction

Between 1925 and 1949, a total of three irrigation engineers were appointed in Nigeria, and all of them were posted to Sokoto. This is probably one of the reasons why Sokoto remained as Irrigation Headquarters up to 1956 even though the Headquarters of the Agriculture Department of Northern Nigeria had shifted from Zaria to Kaduna.

When Mr. John Kitching arrived in Sokoto in June 1949, his activities were concentrated in Sokoto province. Through extensive tours and surveys, he became quite familiar with the situation of irrigation in the province. However, it did not take long before he became involved in irrigation activities in all of the Northern provinces in particular and in Nigeria in general. It was Mr. Kitching who initiated the irrigation policy, first for Sokoto province, and then for Northern Nigeria. What is contained in this policy document is a summary of his writings on the subject between 1949 and 1953; the main policy document was written in 1953. The major sources for this policy are KMAG, File No. 2708, Report Nos. 1 and 2, 1949; File No. 4637/S.I and File No. 26902, 1953; and Nwa (1987) ; and most of the statements are reported as written. Also, it is to be noted that on appointment, Mr. Kitching was advised to focus attention on small to medium size rather than large projects.

5.8.2 *Irrigation Policy for Northern Nigeria, 1953*

1. *Foreword*

Irrigation is not a panacea for all evils which affect farmers in a tropical country. Nor is it always a sign of prosperity, except under conditions where large commercial schemes are feasible. But to an increasing extent irrigation is recognized as a necessity for any tropical country which has a rainfall badly distributed throughout the year, coupled with a rapidly growing population.

2. *Definition of Irrigation*

The chief function of an irrigation engineer is the control of water. Generally, but not always, the purpose of control is to render possible the cultivation of lands which previously were waterless, or waterlogged, or subject to excessive flooding.

3. *Irrigation Policy*

For the time being the general policy in the Northern Region is to construct village irrigation schemes, usually between the limits of 40 to 200 ha. A primary objective of this policy is the collection of data to answer the following questions:

a) Which types of schemes are technically best suited to the various areas?
b) How long will it be before farmers recognize that although to obtain advantage from an irrigation project additional work on their part is necessary, yet the results are sufficiently beneficial to themselves to be worthwhile?

c) Will it prove desirable to concentrate the irrigation effort in provinces where farmers prove to be keen?

4. *Training of Irrigation Staff*

Training of irrigation teams is very important. It is essential to set up a technical organization which is capable of achieving results.

5. *Irrigation Research*

It is also most necessary that at a very early date a research unit should be brought into being, whose function would be to gauge river discharges and collect hydrological data as regards catchment yields and flood runoff.

6. *Method of Irrigation*

So long as abundant surface water remains available for exploitation, irrigation by gravity methods is likely to be a more economical proposition than pumping, which involves the purchase and periodic replacement of expensive plant, and running costs in perpetuity.

7. *Procedure for Village Irrigation Schemes*

The usual procedure adopted for village irrigation schemes is:

i) Reconnaissance by an engineer.
ii) Complete survey by an irrigation team.
iii) Design of the scheme, including detailed working drawings.
iv) Construction, by an irrigation team.
v) Handing the completed scheme over to the Native Administration who then becomes responsible for its operation, annual maintenance and upkeep.

8. *Types of Irrigation Schemes*

a) Dry Season

i) Control of Perennial Streams

The method is to divert a stream so that its water can be used to bring an increased area of land under cultivation. This type of scheme is of limited application as there are relatively few perennial streams. For a time, villagers may also prefer their traditional method of cultivating a fringe of land in the valley bottom, even though the extent of land thus available is very limited.

ii) Storage Dams
The method is to construct a dam across a valley or depression, which impounds flood water during the rains; the stored water being subsequently available for dry season cultivation.

iii) Cattle Dams
(The same principle as for ii). Cattle dams are likely to be immediately popular, because they involve no effort in levelling land, controlling the flow of water, and so on.

iv) Swamp Drainage
The method is to excavate a drain or drains through a permanent swamp so that the stagnant water can flow away. As with cattle dams, little work by farmers is necessary, except in keeping the drainage channels free of bull-rush, etc.

b) *Wet Season*

Flood Protection
The method is to construct an earth dyke or bund to exclude damaging river floods from fadama land. In some cases sluices are provided to admit flood water under control. This type of scheme is superficially attractive, but the technical problems involved are often very considerable.

9. ***Conclusion***

The question may be asked: But is there any real necessity for irrigation work in Nigeria, apart from cattle dams which are of obvious and immediate utility? The answer is that although in many parts of the country, the available farmlands and food crops are sufficient for the needs of the present population, yet the population is increasing so rapidly that in due course there must result a shortage of food. It would be folly to wait for that emergency to arise before remedial measures are initiated, and one remedial measure which to an increasing extent is being adopted in tropical countries is irrigation. The significance of the village schemes now being undertaken is thus not so much their immediate success, as the data they will furnish as a guide to future policy, and the facilities they provide for bringing into being a trained technical staff who are competent to investigate, design and construct irrigation schemes of all types. And it should not be overlooked that each dry season, village scheme constructed enables the local peasantry to improve their diet and raise their standard of living, by working at a time of year when previously they had little occupation.

Chapter 6

Irrigation Development (1960 – 1999)

6.1 General

The period, 1960 - 1999, can really be described as that which saw the rise and fall of irrigation development in Nigeria. The Nigerian Civil War (1967 - 1970) followed barely seven years after the Nigerian Independence in 1960. The Civil War did not initiate but it simply accelerated agricultural stagnation in the war zone that had started after 1965. The continued decline of agriculture was due to its neglect during the period of the oil boom (1970 – 1980).

Because of the deteriorating situation of agriculture, numerous plans and programmes continued to be enunciated all aimed at improving the agricultural situation of the country. These plans and programmes included the 1st - 4th National Development Plans (1962 – 1968), (1970 – 1974), (1975 – 1980) and (1981 – 1985), respectively; the Land Use Decree of 1978; the National Accelerated Food Production Programme of 1973; the Operation Feed the Nation of 1976; and the Green Revolution Programme of 1980 to name a few. Several of these plans and programmes did not have the anticipated positive effect on agriculture in general or on irrigation in particular. The irrigation development between 1960 – 1999 is discussed in the following sections.

6.2 Some Significant Studies on Water Resources and Agricultural Development in Nigeria, 1953-1975

A large number of studies and reports have influenced the development of irrigation in Nigeria in various ways. Obviously all of them cannot be considered here mainly because of space, but then several of them are not easily accessible. Eight of these studies and reports are considered in this section.

In 1953, the Federal Government invited a Dutch engineering firm, NEDECO, and a British firm, Balfour & Company Limited, to undertake investigation and recommend how best to exploit the economic advantages of the Niger and Benue rivers. The Niger Dams Project, which led to the commissioning of Kainji dam in 1968, and Jebba dam in 1983 was the direct outcome of the NEDECO-Balfour studies (NEDECO, 1959). Irrigation was one of the purposes of Kainji dam. In this regard, the immediate impact of the dam was felt by the Nigerian Sugar Company, Bacita. Before the dam, the company could not irrigate its fields before the commencement of the rainy season when River Niger is at its lowest level, but after the construction of the dam, the company could irrigate its cane fields at any time during the year.

In 1963, during the first National Development Plan period (1962-68), the Federal Government requested the FAO "to make an intensive study of the country's agricultural potentials and draw up a scheme of priorities for development" (Nigeria, 1963). In the water resources and irrigation aspect of the study, FAO (1966) reviewed the irrigation situation in the country and then made several recommendations. After stressing the importance of data collection for irrigation development, FAO recommended that data collection should be coordinated and directed by the Federal Government. To coordinate agricultural development at the national level, FAO recommended that a Federal Ministry of Agriculture and Natural Resources be established; one of the departments recommended for this ministry was the Land and Water Use Department which was to assume the responsibility for irrigation and water resources planning and development. This recommendation led to the formation of the Federal Ministry of Agriculture and Natural Resources in 1965 and later to the creation of the Federal Ministry of Water Resources in 1975.

Also in 1963, a cooperative programme was initiated between the Nigerian Government and the United States Agency for International Development (USAID). One of the activities of this programme was to undertake studies that would lead to the identification of specific irrigation projects. The reconnaissance study of Land and Water Resources of Lake Chad Basin, led to the identification of four projects, namely, the Kano River Project, the Hadejia River Project, the Wudil-Gaya Project, and the Takai Project (USDI, 1968). The construction of Tiga dam was recommended and the site for the Challawa Gorge dam was identified. Tiga dam was completed in 1975 for the Kano river Irrigation Project Phase I, the Hadejia Valley Project, and the Kano City water supply, among other purposes and the Challawa Gorge dam was commissioned in 1992 for the same purposes.

Between 1962-68, the UNDP/FAO (1969) study compiled information on the land and water resources of the Sokoto-Rima river valley. The study indicated that there would be enough water to irrigate about 157,000 ha of the upland area and 93,000 ha of the fadama lands. The report recommended that 1,200 ha of pilot project should be developed to introduce and train the farmers for irrigation. The construction of Bakolori and Goronyo dams were recommended. The Bakolori dam was completed in 1978 and Goronyo dam was completed in 1984. The Reconnaissance Report of the Gongola River Project identified the site of the Dadin Kowa dam and 81,000 ha of irrigable land in the area (NEDECO, 1970). The dam was completed in 1988.

In 1970, the National Agricultural Development Committee set up a number of Study-Groups to revise previous reports on agricultural development in Nigeria and prepare for a National Agricultural Seminar which was to consider policies and priorities for development in the period up to 1985. The Study-Group on Irrigation and Drainage (Elahi *et al*; 1971) was one of such groups. The membership of the group represented, as far as possible, every aspect of irrigation and drainage throughout the country. This group was to review the FAO report on agricultural development in Nigeria 1965-80 (FAO, 1966), among others, and to assemble up-to-date data on irrigation and drainage in Nigeria.

The Study-Group reviewed a number of documents dealing with irrigation and drainage in Nigeria and accepted the FAO report as "the basic document for the development of Nigerian Agriculture." In addition to re-emphasizing the importance of data collection to irrigation planning and development, supporting the recommendations in the FAO report, the group made several recommendations of its own. Three of these recommendations were that pilot projects should be established in major irrigation schemes to train both the farmers and others; the states that did not have organizations responsible for irrigation and drainage should do so; and a National Water Resources Institute should be established in the country. Most of these recommendations have since been implemented. For example, the National Water Resources Institute started operating in 1978 in Kaduna, but received legal recognition in 1985.

The Hadejia River Basin Study (Schultz International Limited, 1975) evaluated the land and water resources in the Hadejia basin and found that, excluding the Kano River Project and its extended project area which had been identified previously, approximately 190,000 ha of land in the basin was suitable for irrigation. However, the water in the Hadejia river was not enough to support both the Kano River Extended Project Area and the Hadejia Valley Projects as proposed at the time. This study recommended that a development authority should be established for the Hadejia Jama'are River Basin as had been done for the Chad and the Sokoto-Rima Basins. It also recommended that a federal statute should be enacted to transfer the right to regulate water to the Federal Government as provided under the Constitution. The decree that set up the Hadejia-Jama'are River Basin Development Authority, among others, was promulgated in 1976 and the Water Resources decree in 1993.

All the above documents, among others, contributed immensely to the Third National Development Plan document, 1975-80 (Nigeria, 1975) which included a coherent irrigation development plan for Nigeria.

6.3 Dams for Irrigation

6.3.1 Introduction

The earliest recorded attempt to construct a modern dam in Nigeria was made by the Public Works Department in 1912 at Kwa Falls to provide water and electricity for the emerging economic and political capital city of Calabar in the present Cross River State. The project was eventually abandoned but success was achieved by the Nigerian Electricity Supply Company (NESCO) when they completed the first modern Nigerian dam in 1923 at a sight very similar to the one above, Kwall Falls, near Jos the capital of the present day Plateau State (Sadiq, 1994).

Probably, irrigation was not one of the objectives of the 1923 dam but the present dam at the same site under the Lower Benue River Basin Development Authority has irrigation as its priority objective.

For discussion purposes, the dams have been divided into three groups, large, medium and small according to their height, h. The groups are as follows:

a) large dams: $h \geq 30$ m
b) medium dams: $15 \text{ m} \leq h < 30$ m
c) small dams: $h < 15$ m

6.3.2 Existing Small Dams for Irrigation

One of the oldest irrigation dams in Nigeria is the Wurno dam built in 1960 by the Public Works Departments for the Wurno Irrigation Project in Sokoto State. This was quickly followed by a small irrigation dam in Samaru, near Zaria, the Bomo dam, which was planned in 1960 and constructed in 1961 by the Station Staff. At that time, Samaru was the Headquarters of the Regional Research Station of the Northern Nigeria

Ministry of Agriculture. The dam was built across Bomo stream to provide surface and sprinkler irrigation to some of the farms at the Research Station.

The number of existing small irrigation dams in Nigeria was put at 42 in 1992 (NWRMP, 1995). This must be a rough estimate because a number of small dams were excluded and also a number of others that were planned but not constructed were included in the list. Table A.1 in the appendix gives the existing small irrigation dams in 1992. This table was compiled from several documents including the national inventory on dams by the Sub-Committee on Dams of the National Technical Committee on Water Resources, published documents by project owners, and the Sector report of the National Water Resources Master Plan (NWRMP, 1995), among others. Some details of several small dams were not available.

6.3.3. Existing Large and Medium Dams for Irrigation

The oldest large dam and the largest dam in Nigeria to date is the Kainji dam in Niger State which was commissioned in 1968. Although hydroelectric power was the principal objective of the dam, irrigation was certainly one of its objectives from inception as provided for in the Niger Dam Authority Act of 1962 and the NEPA Decree of 1972. Currently water is being pumped from Kainji lake to irrigate small-and medium-schemes around the lake but the largest irrigation scheme to benefit from the dam so far is the Nigerian Sugar Company Estate at Bacita. The dam has made it possible for the sugar company to irrigate its 5,960 ha of sugar cane estate without any difficulties with water.

The second large to medium size irrigation dam in Nigeria is the Bagauda dam which was completed by the Kano State Ministry of Works in 1970. The dam was built to irrigate approximately 600 ha of land and provide 114,000 m^3 of water per day for water supply in addition to providing facilities for fisheries and recreation.

As a result of a heavy rain in August 1988, Bagauda dam failed; the estimated number of houses destroyed and the number of people left homeless were 18,000 and 200,000, respectively; 23 lives were claimed to be lost while 2 persons were declared missing by the State Government. The dam was rehabilitated the same year.

The number of existing large and medium irrigation dams in Nigeria was put at 29 in 1992 (NWRMP, 1995). The details of some of these dams are given in Table A.2 in the appendix.

6.4 Small-Scale Irrigation Schemes

Irrigation systems have been classified in a number of ways. Sometimes, the level of technology and type of management have been combined with size to classify an irrigation system. In this publication, the scheme size is taken as the basis of classification. There has been some discussion whether "a large-scale scheme" with many farmers in which one farmer or one farm-family has a maximum of say 4 ha is really large-scale or small-scale. In this publication such a scheme is classified as large or small-scale once it meets the size criterion irrespective of the number of farmers on the scheme or the size of individual holding.

The size category adopted in the National Water Resources Master Plan (NWRMP, 1995) is also adopted here. Three size categories are recognized as follows:

a) large-scale : schemes > 1,000 ha
b) medium-scale : 500 ha < schemes > 1,000 ha
c) small-scale : schemes < 500 ha.

Between 1960 and 1970, emphasis continued to be placed on the development of small-scale irrigation schemes in Nigeria most of which were to be found in the Northern Region. While trying to consolidate the schemes that were already developed before this period, a number of new developments were undertaken. Schemes such as Gunuar Kuka, Arawa and Mai-alkama in the present-day Jigawa State, were developed in 1960 for wheat production, pumping water from the Hadejia river. Ta Hoss scheme in Plateau State was first irrigated in 1962 from a natural spring near the railway line which provided the water to irrigate maize, rice and vegetables.

In 1963, 100 ha of land was developed in Bakura near Sokoto, to irrigate 50 ha of rice and 50 ha of orchard from Lake Natu while less than 20 ha was developed at Wawan Rafi in Kano State to irrigate vegetables. Pumping water from Ebeji river for wheat production, the Gamboru scheme in Borno State was first irrigated in 1964. An interesting project in this area was the North Chad scheme which proposed to put 40 ha of land under rice and wheat, pumping water directly from Lake Chad. This scheme was first irrigated in 1966 but was later abandoned completely because Lake Chad receded away from the project site due to a series of droughts (Iqbal, 1975). This was similar to the situation reported by Nwa (1993) when Lake Chad was said to have receded by up to 60 km due to the 1984-85 drought.

In 1953, John Kitching an irrigation engineer stationed in Sokoto but with an extensive irrigation experience in Nigeria (KMAG, File No.

4637/S.I; 1953) stated that "no irrigation projects have yet been undertaken in either the Eastern or Western Region." Before 1960, only 60 ha of land was under controlled irrigation in the Eastern Region and that was for rice irrigation. The first and the only irrigation engineer in the East at that time was recruited in 1963. He was attached to the Division of Agriculture and was responsible to that division's Adviser on rice promotion. Except for the Uzo-Uwani irrigation scheme, the construction of which started in 1963, and which is discussed under medium and large-scale irrigation schemes, irrigation development activities in the Eastern Region up to the beginning of the civil war in 1967, were limited to providing irrigation facilities to small areas varying in size from 0.40 to 16 ha.

In Western Region the situation remained almost the same up to 1970 as noted by John Kitching in 1953. However, in 1973, the Head of State of Nigeria, General Yakubu Gowon, visited the Peoples' Republic of China, and noting the extent of rice cultivation in that country, invited the Chinese to come to Nigeria and assist in training Nigerians in rice cultivation. Itoikin, near Lagos, was one of the three locations selected by the Chinese Team for rice cultivation in the country. Out of the 1,300 ha of land acquired for the project, only 315 ha was later found to be suitable for agricultural development out of which only 141 ha was found suitable for irrigation; all of the 141 ha of land was actually under rice, maize and vegetables irrigation in 1992. The problem of Itoikin was that the initial choice of the site was not backed-up by a feasibility study. Itoikin scheme was transferred to Ogun-Oshun River Basin Development Authority in 1978.

The second location selected by the Chinese Team was Shonga in Kwara State. It appears that the Chinese did not start any work at Shonga. In 1987 only 11 ha of land was irrigated at the site. Shonga scheme now belongs to the Lower Niger River Basin Development Authority. The third location selected by the Chinese was Odekpe in Anambra State where, like Shonga, no development was started at that time. The Anambra-Imo River Basin Development Authority took over Odekpe scheme when the Authority came into existence in 1976.

Although irrigation development, including small-scale schemes, has been very slow in Nigeria, there is hardly any state in the federation today that does not boast of one or two small-scale irrigation scheme(s). Table A.3 in the appendix gives information on the existing small-scale irrigation schemes at various stages of development in the states.

6.5 Water Resources and Engineering Construction Agency (WRECA) of Kano State

The history of irrigation and water resources development in Nigeria cannot be complete without a mention of the Water Resources and Engineering Construction Agency (WRECA) of Kano State. This is so because WRECA has built more dams and developed more irrigation facilities than any other agency in Nigeria apart from the Federal Government itself. Nearly all the damable sites in Kano State have been dammed by WRECA. Figure 6.1 gives a pictorial summary of the activities of WRECA in Kano State between 1970-1985. Several of the dams shown in the figure as being proposed or under construction between 1970-85 had long since been constructed. The details of these dams are given in appendices A1 and A2.

Kano was one of the states created in Nigeria by the Head of State, Lt. Colonel Yakubu Gowon in May 1967 at which time the Kano State Ministry of Works and Surveys was set up. The late Eng. Balarabe Ismaila was the permanent secretary of this ministry between 1968 and 1969 and was succeeded by Alhaji Salihi Iliyasu between 1970 and 1973. The Water Resources Division of the ministry was responsible for the major water resources activities in the State. For example, this division started the construction of Tiga dam in 1971.

In 1974, the Kano State Government under the Military Governor, Police Commissioner Audu Bako, decided to separate the Water Resources functions from the Ministry of Works and Surveys and accordingly set up a new organization called "Water Resources and Engineering Construction Agency (WRECA)" to take over those functions (Haskoning, 1978). The construction of Tiga dam that was commenced by the Water Resources Division of the Ministry of Works and Surveys in 1971 was completed by WRECA in 1975. WRECA has been managed by the following Managing Directors:

- Sept 1974 - Oct 1974 Mr. V.S. Kalatunga
- Oct 1974 - Nov 1986 Alhaji Magaji Abdullahi
- Dec 1986 - Aug 1988 Alhaji Abdulaziz Umar
- Aug 1988 - May 1990 Alhaji Garba Usman
- May 1990 - Alhaji Bello Muhammed

In 1990, WRECA was split into two, namely, the Kano State Water Board, whose responsibility is to manage and maintain water supply schemes, and the new WRECA which was mandated to undertake water resources development, building and civil engineering contracts within and outside the state (WRECA, n.d).

6.6 National Fadama Development Project (NFDP): The Evolution of an Idea

6.6.1 Introduction

In 1992, the Federal Government of Nigeria signed an agreement with the World Bank to provide a loan to finance a project for developing fadama lands by introducing small-scale irrigation. The project started in five states, i.e., Bauchi, Jigawa, Kano, Kebbi, and Sokoto and slowly went on to eventually cover the 36 states of the Federation and the Federal Capital Territory of Abuja. The project, known as the *National Fadama Development Project (NFDP),* aimed at accelerating the pace of fadama development by:

- improving infrastructure into and within the fadama;
- using improved but simple technology to develop shallow tubewells;
- privatizing drilling activities and building the capacity of the state and federal institutions to support fadama development;
- improving irrigation management through water users' associations;
- conducting aquifer studies and monitoring and upgrading irrigation technologies; and
- undertaking environmental assessment for further expansion of irrigated agriculture (World Bank, 1992).

On one hand, the use of the fadama land in Nigeria has gone on for centuries while the study of these lands and research into their use have gone on for decades. On the other hand, the Nigerian governments, including the colonial government, have struggled over the years with ideas and programmes that could make Nigeria self-sufficient in food and fibre; that objective is yet to be achieved. NFDP is another of such ideas and programmes. The purpose of this Section (6.6) is to explore how the idea of the "shaduf in the fadama" evolved into the "National Fadama Development Project (NFDP)."

6.6.2 Shaduf in the Fadama

As indicated in Chapter 2, the shaduf probably moved into Northern Nigeria from Egypt or other parts of North Africa in about AD 700.

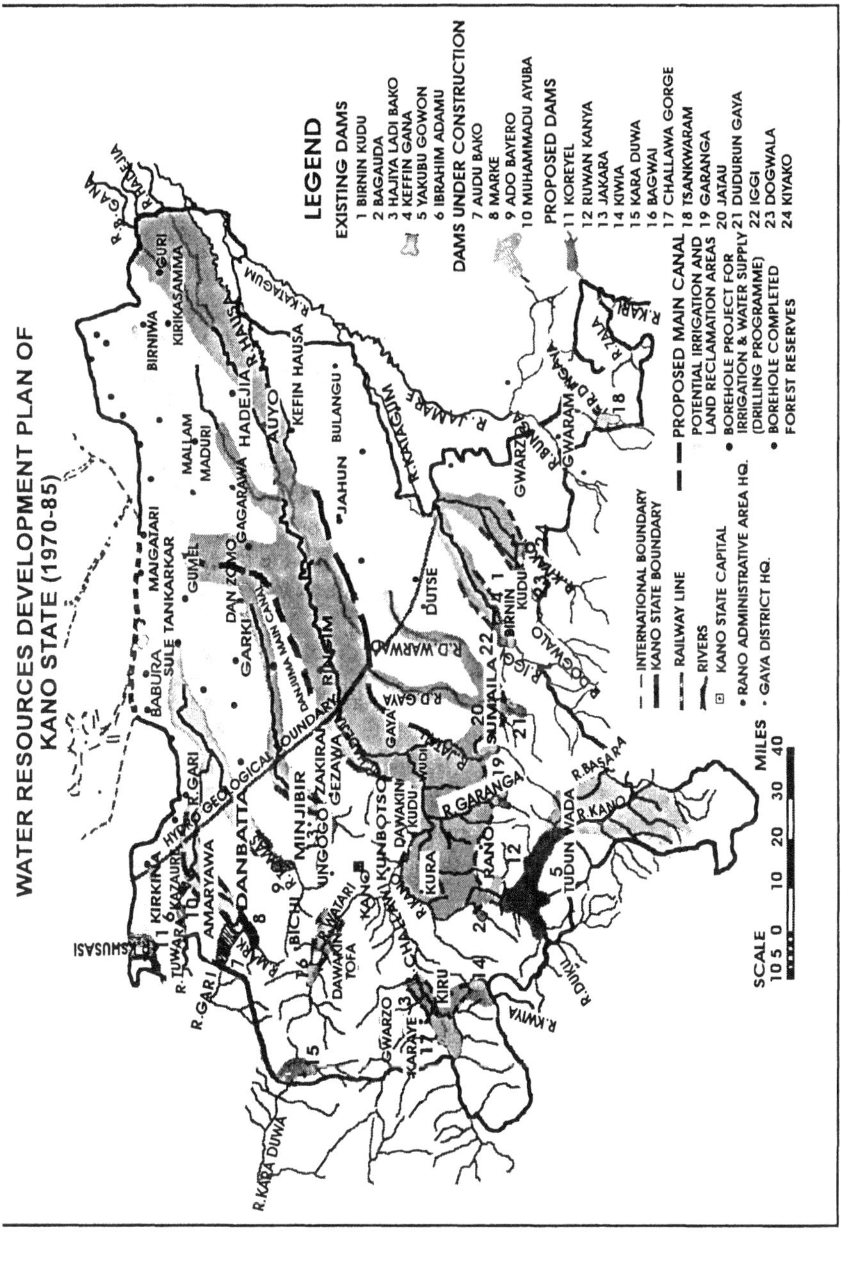

Fig. 6.1 Summary of the activities of WRECA, 1970 - 1985

Since then, the shaduf has been operated in all parts of Northern Nigeria till today. During his survey of conditions for irrigation in Nigeria in 1944, an irrigation engineer based in Sierra Leone, Mr. C.J. Rae, saw the extensive fadama lands all over Northern Nigeria and shaduf irrigation in most parts of the Region. When he visited Kano, he saw the oxbows of the Jakaradi river, the intensively cultivated areas along the Challawa river, and the extensive vegetable gardens on the northern fringe of Kano City, all irrigated with the shaduf. He wondered why small pumps or animal driven water lift system, which would be less tedious than the shaduf, were not tried. He suggested that electrically driven pumps could be tried especially in areas close to Kano City while the animal driven lift system could be tried elsewhere to replace the shaduf. In 1965, Mortimore and Wilson (1965) found that, in the Kano area, one person could lift, with the shaduf, 1.95 litres of water per second; the irrigation rate was 0.75 l/s/ha while the interval of irrigation was 2-7 days. It appears that Mr. Rae's suggestions were not taken up immediately and directly. Small petrol engine and diesel pumps (not electrical) were introduced later into the fadama lands in a big way but the use of animal driven lift systems was very limited.

6.6.3 Shaduf vs Small Pump Studies

It is not easy to determine exactly when small irrigation pumps were introduced into the fadama. Erhabor (1982) suggested that the pump was first introduced to the farmers in Nigeria by agricultural instructors and friends of the farmers in the early 1970s. Probably the late 1960s may be closer to the mark because in 1971 a fairly large number of farmers were already using the pumps in the Zaria area and it is not likely that all those pumps would have been introduced within one or two years.

Probably the earliest project in Nigeria that officially assisted the farmers to obtain irrigation pumps was the Cadbury tomato project. In 1971, Cadbury Nigeria Limited together with the Kaduna State Government set up a tomato paste factory in Zaria. The local farmers were assisted with inputs, including small pumps, and extension service, and in return they were to sell their tomato to the factory (Quinn, 1971; Agbonifo, 1974). This project collapsed five years after it was set up. One of the earliest research work that had shaduf and small pump components was conducted to find out some of the reasons for the collapse of the Cadbury tomato project. During the 1975-76 dry season, Orewa (1978, 1981) conducted a study to find out the technical and management problems and the economics of the Cadbury tomato project. One of the findings of the study on the economic side was that the production costs were greater with the shaduf system than with the pump system. Most of the farmers hired their pumps. On the technical side, one of the greatest problems of the project was the frequent breakdown of the irrigation pumps. Closely following this problem and directly related to it was the shortage of trained technicians to maintain the pumps. These were some of the technical problems that contributed to the collapse of the Cadbury project.

In her study of the classification, spatial variation, present and potential use of the fadama lands of central Northern Nigeria, Turner (1977) listed the advantages and disadvantages of the shaduf and the small pump in fadama irrigated farming. The following were listed for the shaduf:

- one shaduf can irrigate not much more than 0.10 ha;
- the area of land irrigated must be adjacent to the water supply; the conditions on a terrace may be ideal for agriculture while the fadama may remain waterlogged;
- the maximum height of lift for the shaduf is about 3 - 4 m;
- the amount of capital needed is small; and
- the amount of labour involved is considerable.

With the pump:

- one pump can irrigate up to about 4 ha;
- water can be lifted to a much greater height or to an area not immediately adjacent to the water supply; conditions on a terrace may be ideal for agriculture while fadama may remain waterlogged; and
- labour requirements are much lower.

Between 1977 and 1979, Nwa (1981a) conducted a detailed study during two irrigation cropping seasons, to evaluate small pump and shaduf systems of irrigation in the Zaria area. Before this time, the pumps used by the majority of farmers were either hired or highly subsidized. The main objective of the study was to find out whether a small farmer could easily recover an investment made by purchasing his own pump. The conclusion of the study was that a shaduf farmer could, with a high value crop, recover the investment on a small pump within a short period of time (1, 2 or 3 crops, not cropping seasons). However, it was also strongly suggested that shaduf farmers or other farmers could be encouraged to buy their own pumps in a massive way only when credit and pump service arrangements must have been assured.

After the first year of the study by Nwa (1981a) in the Zaria area, precisely between October 1978 and May 1979, Erhabor (1982) undertook his study in Ringim Local Government Area of Kano State to estimate the economic returns to small-scale shaduf and pump irrigation systems. The results of the study showed that the returns to shaduf and pump irrigation farming were quite high. It was also noted that lack of timely repair service was what hindered the adoption of pump irrigation at that time.

6.6.4 Pumps and Tubewells in the River Basin Authority

Before the ADPs got variously involved in the use of washbores and tubewells in the fadama, some river basin authority have previously accumulated some experience in that aspect of work. For example, in the mid-70s the Chad Basin Development Authority (CBDA) used washbores and tubewells connected to 2" or 3" petrol driven water pumps to exploit shallow acquifers for irrigation in the fadama areas of Jare Bowl and around Maiduguri under the leadership of Mr. Kamar Omar.

6.6.5 ADPs and their Supervising Organizations

While the small farmer was struggling with his shaduf in the fadama to produce more food, and while researchers were trying to show that the shaduf farmer could lighten his burden with a small pump and still make good returns on his investment, the Federal Government was struggling with policies and programmes that could accelerate food production in the country.

Nigeria is said to have experienced a food problem perhaps for the first time during the Second World War (1939-1945) when agricultural activities declined due to war disruptions. The problem abated for sometime thereafter but reappeared between 1960-1970 (FMANR, 1984). From 1970, agricultural production stagnated with annual growth rate of less than 1 per cent, domestic food prices rose, agricultural exports declined (World Bank, 1982). The situation has continued to deteriorate to date (1999). At first, this problem was solved by food importation (Abalu and D'Silva, 1980) but because of the declining ability of the government to continue with the importation, attempts were made and are still being made to solve the problem through various policies and programmes.

One of the several programmes adopted by the government to solve the food problem was the Agricultural Development Project (ADP) which aimed at increasing agricultural production by means of improved farmer services to include extension, inputs, credit, marketing, rural roads and water supplies, among others. With loans from the World Bank (1977), the Federal Government launched in 1975 three enclave Integrated Agricultural Development Projects (IADPS) at Gombe in Bauchi State, Gusau in Sokoto State and Funtua in Katsina State. These were quickly followed by other enclaves, Lafia (Plateau State) and Ayangba (Kogi State) in 1977, Bida (Niger State) and Ilorin (Kwara State) in 1979, and Oyo North (Oyo State) and Ekiti-Akoko (Ondo State) in 1980.

In 1981, Gombe ADP became statewide and, therefore, Gombe became the first state to undertake a statewide ADP. Kano State (now Kano and Jigawa) followed Bauchi later in the same year, 1981, tó undertake a statewide ADP. Gusau ADP became statewide in 1982 and, therefore, Sokoto State (now Zamfara) became the third state to undertake a statewide ADP. Thus, the five states of Bauchi, Jigawa, Kano, Kebbi and Sokoto led the way in undertaking statewide ADPs. The ADPs were later expanded to cover all the 36 states and the Federal Capital Territory, Abuja.

With the assistance of the World Bank, the Agricultural Projects Monitoring, Evaluation and Planning Unit (APMEPU) was established in 1974 to coordinate the ADPs. In 1976, the Federal Department of Rural Development (FDRD) was created within the Federal Ministry of Agriculture and Natural Resources (FMANR). The FDRD took over the functions of evaluation and monitoring of the ADPs although these responsibilities were still carried out through APMEPU. Because of the rapid expansion of the ADPs, a new organization, the Federal Agricultural Coordinating Unit (FACU) was created in 1981 also with the assistance of the World Bank to take over the coordination of the ADPs. FACU itself was also supervised by the FDRD. FACU was responsible for the identification, preparation, appraisal and successful take-off of the new statewide ADPs as well as the technical support and supervision of all ADPs. When FACU was created, the activities of APMEPU were limited to monitoring and evaluation and, therefore, the name of the unit was modified to Agricultural Projects Monitoring and Evaluation Unit (APMEU).

To implement the agricultural development projects, many of the states, especially the five leading ADP states, quickly established Agricultural Development Authorities. For example Kano Agricultural and Rural Development Authority (KNARDA) was established by the Kano State Government in 1982. This was quickly followed by Sokoto Agricultural and Rural Development Authority (SARDA) and Bauchi State Integrated Rural Development Authority (BASIRDA). The Kebbi Agricultural and Rural Development Authority (KARDA) and Jigawa Agricultural and Rural Development Authority (JARDA) were established in 1992 at the time they were preparing to obtain their NFDP loan.

6.6.6 Pumps and Tubewells in the ADPs

As a follow-up of the work done on pumps in the fadama by researchers, some ADPs carried out some work in their projects but this time to test and identify the specific type of pumps that were most suitable to their areas. In 1982, KNARDA tested several diesel and petrol pumps as well as a solar pump, and some of the diesel pumps were recommended for use in the area. The solar pump was not recommended because of insufficient sunlight during the harmattan (the irrigation season) and the prohibitively high prices for the system (Brown, 1982). Hand pumps were also tested and used but the farmers soon lost interest in them because they were not much better than the shaduf especially regarding the effort (human energy) required to operate them. Meanwhile, some of the ADPs, especially Bauchi, had invited Indian consultants to introduce the tubewell technology into the fadama. The technologies were largely the shallow tubewell by washboring (washbore, no drilling rigs are required) and the shallow tubewell by drilling (drilling rigs are used). In 1982, installation of tubewells was initiated in the Jama'are fadama of Bauchi State. Kano State followed with its own trials almost immediately after Bauchi while SARDA first conducted field trials on tubewells at Bunza and Wurno in Sokoto State in 1984. Meanwhile, farm surveys and studies regarding the viability of small pumps and tubewells in the fadama continued especially within Bauchi and Kano ADPs. For example, the studies and surveys of Chapman in Bauchi strengthened the existing views that small pumps and tubewells in the fadama were highly economically viable (Chapman, 1983; 1984).

In 1983, Kano State imported 2,100 petrol engine pumps (51 mm and 76 mm sizes) and sold them to the farmers at subsidized rates; in the same year the state had two small drilling rigs operating, and had drilled 35 successful tubewells. Also in 1983, Bauchi State sold 380 small pumps and drilled 79 successful tubewells.

6.6.7 NFDP

Bida enclave ADP in Niger State was the first to introduce irrigation into the ADP. They used controlled flooding by constructing water control structures, head dykes and two peripheral canals with a drain in the middle. This system permitted the cultivation of rice and some dry season vegetables (Mijindadi *et al.*, 1993). Bauchi, Kano and Sokoto ADPs started fadama irrigation by upland water conservation. Bunds were used to impound runoff water during the wet season and the residual moisture was used to produce a second crop; this did not achieve the expected results. It was after this unsuccessful method of fadama irrigation that the three ADPs of Bauchi, Kano and Sokoto started their fadama irrigation programmes as indicated in the previous sections.

Based on the experiences of the 3 ADPs (Bauchi, Kano and Sokoto), FACU developed the proposal for NFDP which led to the signing of a loan agreement between the World Bank and the Federal Government to start the project in 1992. As indicated previously, NFDP started initially with the 5 states of Bauchi, Jigawa, Kano, Kebbi and Sokoto; the project has already spread to several other states and there are plans to cover eventually all the states of the federation. Table 6.1 shows the tremendous effort made by the ADPs to develop the fadamas even before NFDP became operational. The result should be much more striking when NFDP becomes effective in all the states.

6.7 Large and Medium-Scale Irrigation Schemes

As could be seen in Chapter 5, experience in irrigation was building up over the years; perhaps it was inevitable that Nigeria would sooner or later start to develop medium- and large-scale irrigation schemes. Between 1960 and 1970, reconnaissance, pre-feasibility and some feasibility studies had been conducted in the basins of Cross river, Gongola river, Kano river and Lake Chad, and in the valleys of Hadejia, Sokoto-Rima and Niger rivers. In addition, some pilot irrigation projects had been started in some of the above areas. The leading irrigation schemes in this period were the Komadugu-Yobe irrigation project comprising Yau scheme which was started in 1960, Daya scheme started in 1961 and Abadam scheme which was started in 1964, all in Borno State; the Ebeji scheme also in Borno State was started in 1963, and also started in 1963 was the Bacita Sugar Estate in Kwara State.

*Table 6.1: Irrigated area (ha) through washbores, tubewells, direct pumping and other methods by some State ADPs (1985-1992)**

State	Washbores tubewells	Direct pumping	Diversion, flood control and methods	Total
Adamawa and Taraba	200	1,300	2,000	3,500
Bauchi	5,200	17,000	1,000	23,200
Benue	20	2,000	500	2,250
Borno and Yobe	1,550	17,500	3,050	22,100
Kaduna	400	13,000	0	13,400
Kano and Jigawa	5,700	56,000	4,000	65,700
Katsina	250	500	0	750
Kwara	100	400	0	500
Niger	10	2,500	5,950	8,460
Plateau	290	11,800	1,000	13,000
Sokoto and Kebbi	6,300	19,000	500	25,800
Total	**20,020**	**141,000**	**18,000**	**179,020**

Source: Mijindadi et al; 1993
**Data prior to the creation of new states in 1991, that is why two states are given in some cases.*

The end of the Nigerian Civil War in 1970 was also the beginning of the "oil boom" period in the country. The availability of "cash" to the Federal Government definitely encouraged the heavy investment in irrigation during that period. The total capital expenditure on irrigation in Nigeria between 1970-1984 was about 2.34 billion (2.34×10^9) Naira (Adeniji, 1985). At 1979 Naira value, these costs were estimated to range from ₦6,000 to ₦12,000 per hectare (in January 1980 1.00 = US1.75 = £0.81). It was during this period that most of the medium and large-scale irrigation projects were established in the country. The summary of some of the information on the large and medium-scale irrigation schemes are given in Table A.4 in the appendix while some of the projects are described in the following sections.

6.7.1 Nigerian Sugar Company Estate, Bacita

It was in 1956 that the Nigerian Government invited Bookers McConnell Limited, a British sugar manufacturing company to investigate the feasibility of establishing a sugar industry in the country. In 1958, the company submitted a report suggesting Bacita as the site for the sugar estate. The report was accepted by the government which set up a "Sugar Syndicate" to further investigate the production of sugar in the area recommended. The syndicate recommended the formation of the "Nigerian Sugar Company Limited" which was accepted by government and the company was incorporated in 1961. Bookers (Nigeria) Limited, a subsidiary of Bookers McConnel was appointed the Managing Agents for the company in 1963 (NSC, n.d.).

Bacita (9°N and 5°E) is a small town on River Niger about 120 km north of Ilorin, the Kwara State capital. The irrigation water for the sugar estate is obtained by diverting water above a weir constructed across the River Niger into a canal which transports this water to the estate. The water is lifted at a pumping station into the main canal that carries the water to the irrigation areas. The estate is divided into two areas, one area is irrigated by the sprinkler method (hand move) and the other area is irrigated by the surface method (furrow). Water is also pumped directly from the river to irrigate parts of the sprinkler area. In 1980, a total of 4,954 ha of land was irrigated out of which 2,788 ha was under sprinkler and 2,176 ha was under surface irrigation. Figure 6.2 shows the layout of these areas relative to River Niger.

Makanjuola and Nwa (1985) showed that there was a general decrease in cane yields over the years at the estate especially between 1974 and 1982 (Tables 6.2 and 6.3). This was attributed to the application of excessive water over the years together with the associated general rise in the groundwater table. Improved water management which includes, irrigation schedules that take the crop stage of growth into consideration, proper irrigation stream size and furrow length, and adequate land drainage, were suggested as actions that would contribute to the reversal of the declining yield trends.

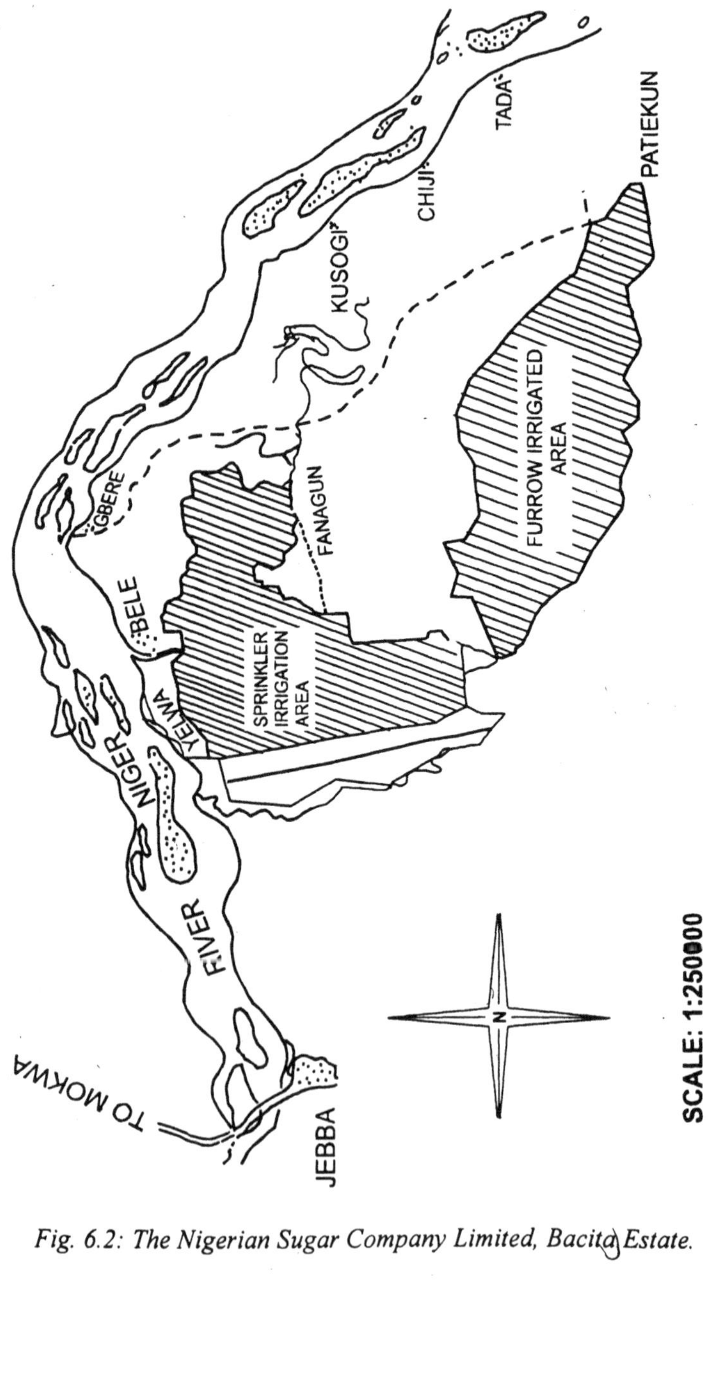

Fig. 6.2: The Nigerian Sugar Company Limited, Bacita Estate.

Table 6.2: Annual water balance and yield for surface irrigated area.

Year	Yield (t/ha) (cm)	Total water applied*	Crop water requirement ETcrop (cm)	Water balance++ (cm)	Cumulative water excess applied (cm)
1974	74	191.5	153.3	+ 38.2	38.2
1975	77	172.6	162.3	+ 10.3	48.5
1976	66	187.9	157.4	+ 30.5	79.0
1977	71	156.6	161.2	- 4.6	74.4
1978	63	141.2	151.7	- 10.5	63.9
1979	73	166.1	155.7	+10.4	74.3
1980	67	157.6	165.5	- 7.9	66.4
1981	58	178.2	168.5	- 9.7	76.1
1982	59	151.4	167.5	- 16.1	60.0

* *Sum of Irrigation Water and Rainfall*
++ *Column 3 - column 4, (+ve) indicates excess while (-ve) indicates deficit*

Table 6.3: Annual water balance and yield for sprinkler irrigated area.

Year	Yield (t/ha)	Total water applied* (cm)	Crop water requirement ETcrop (cm)	Water balance++ (cm)	Cumulative water excess applied (cm)
1974	85	205.3	153.3	+ 52.02	52.0
1975	85	198.2	162.3	+ 35.9	87.9
1976	78	214.0	157.4	+ 56.6	144.5
1977	72	184.2	161.2	+ 23.0	167.5
1978	72	152.6	151.7	+ 0.9	168.4
1979	67	183.8	155.7	+ 28.1	196.5
1980	66	172.1	165.5	+ 6.6	203.1
1981	66	187.9	168.5	+ 19.4	222.5
1982	52	160.2	167.5	- 7.3	215.2

* *Sum of Irrigation Water and Rainfall*
++ *Column 3 - column 4, (+ve) indicates excess while (-ve) indicates deficit.*

6.7.2 Uzo-Uwani Irrigation Scheme

In 1961, the Eastern Nigeria Government, acquired 4,270 ha of land in Uzo-Uwani, 58 km South of Nsukka, in Do-Anambra river basin, in the present-day Enugu State to establish a farm settlement scheme. Irrigation of two crops of rice per year was proposed on 1,000 ha of land. The land was re-distributed to selected settlers who obtained 6 ha of irrigation land and 1.2 ha of dry-crop land. The construction work was started in 1963. A stone paved earth weir across the Obina river raised the water level such that the water flowed by gravity into the 8 km long canal that carried the water to the irrigation area. In 1980, 315 ha of the scheme was irrigated.

6.7.3 Kano River Irrigation Project (KRIP), Phase I

As indicated in section 6.2, one of the projects identified in the reconnaissance study of Land and Water Resources of Lake Chad Basin was the Kano River Irrigation Project (KRIP). KRIP Phase I covers an area of 22,000 ha and lies between latitudes, 11° 30'N, 12° 03'N and longitudes 8° 30'E, 9° 40'E within the Hadejia-Jama'are River Basin, approximately 30 km South East of Kano City on both sides of Kano-Zaria Express Road (Figure 6.3, Nwa, 1982). By 1994, 15,000 ha of the project had been equipped and irrigated. Initially, KRIP was managed by the Kano State Ministry of Agriculture and Natural Resources, but it was passed to a Federal Agency, the Hadejia - Jama'are River Basin Development Authority (HJRBDA), when it was created in 1976 (Nwa and Martins, 1982).

The water for the project is conveyed from Tiga dam to the irrigation area through a 12 km long main canal which splits into the East Branch canal (27 km long) to serve 14,000 ha and West Branch canal (32 km long) to serve 8,000 ha. In 1992, the project had 53 km of secondary canals, 174 km of tertiary canals, 1120 km of quaternary canals (field channels), and 9 night storage reservoirs. The night storage reservoirs receive and store the flow from the branch canals during the night for irrigation during the day. This arrangement permitted the design of smaller branch canals otherwise larger size canals would have been required to carry enough water for daytime irrigation.

The entire project operates the gravity irrigation system with wheat and tomato as the main initially irrigated crops; rice is grown during the wet season with supplementary irrigation. The average land holding in the project is 0.4 ha for a family of six (Ben-Musa, 1992).

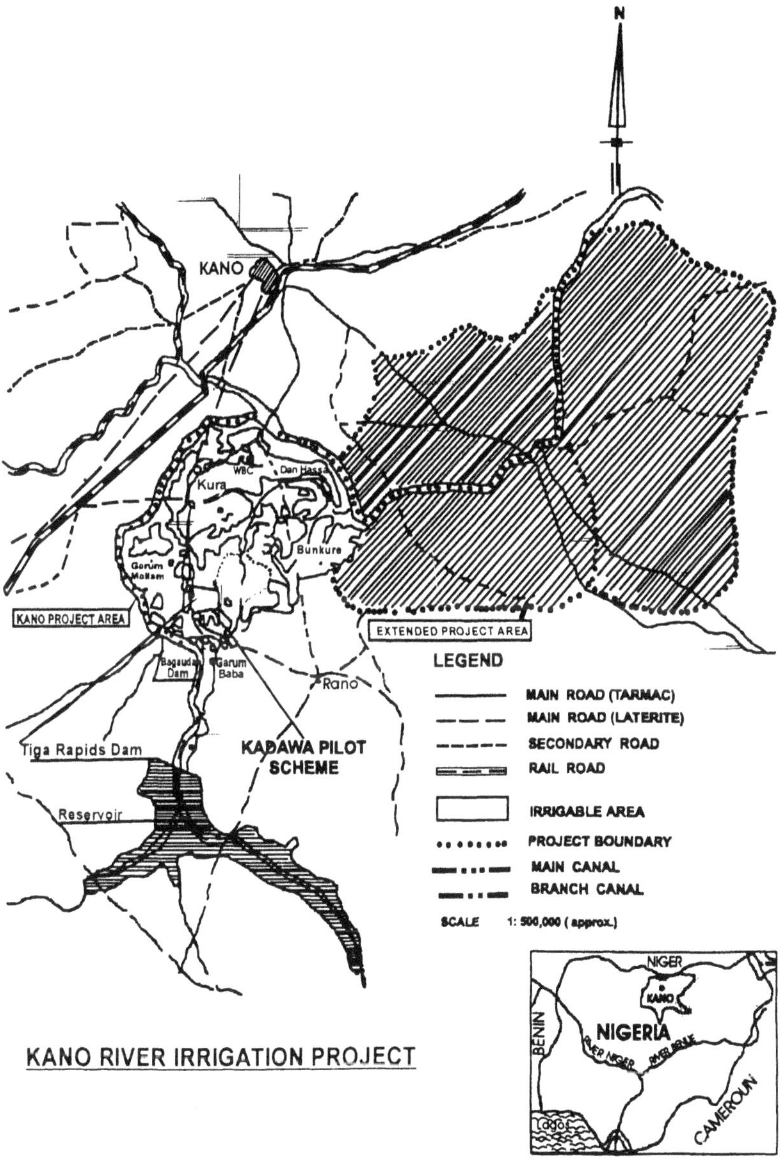

Fig. 6.3: Kano River Irrigation Project, Phase 1

Water logging and rising groundwater table were observed by Nwa (1982) in parts of the project area, especially in the Kadawa sector. Groundwater table in the area was said to be generally below 1.5 m of the soil surface before the introduction of the project. After the introduction of the project, water table was found to exist in parts of the area between 0 and 60 cm below the ground surface at the peak of the rains and at 1 m or less during the peak of the irrigation season. This situation was attributed to over irrigation, high rate of seepage from the unlined canals and field channels and the poor maintenance of the surface drainage system (there is no sub-surface drainage system).

6.7.4 South Chad Irrigation Project (SCIP)

The pilot study of the South Chad Irrigation Project (SCIP) was started in 1970 when the Commonwealth Development Corporation (CDC) started to investigate the large scale production of wheat and rice in the area through small holders and mechanized means. In 1971 the feasibility study of the project was commissioned by the North East State Government with UNDP/FAO assistance. The Federal Government took charge of the South Chad Irrigation Project and water resources development and agricultural production in the area, and formally established the Chad Basin Development Authority (CBDA) by Decree No. 32 in 1973 (Nwa and Martins, 1982).

SCIP, situated about 120 km north-east of Maiduguri, was proposed to cover an area of 67,000 ha to be developed in three stages. Stage I (Central area) was to cover 22,000 ha, Stage II (Eastern area) was to cover 27,000 ha, and Stage III (Western area) was to cover 18,000 ha. Designed to take irrigation water from Lake Chad, construction of the scheme started in 1975. An intake channel, 38 km long, conveys water from the lake to the main pumping station with a pumping capacity of 85 m^3/s at Kirenowa, where the water is lifted 10 m into a 20 km long main canal. The main canal conveys the water to the Gadadai pumping station where it is lifted 6 m by three 30 m^3/s capacity pumps into a distribution basin (Figure 6.4) from which water is distributed to the various irrigation areas through a system of branch canals and distributaries.

Part of Stage I of the project was commissioned in 1979 when the first irrigation took place; in 1992, 7,000 ha of this Stage was irrigated. Each farm family in the project is provided with 4 ha of irrigation land. Wheat was the original main dry season crop while vegetables were also cultivated; rice was the main wet season crop while cotton, maize,

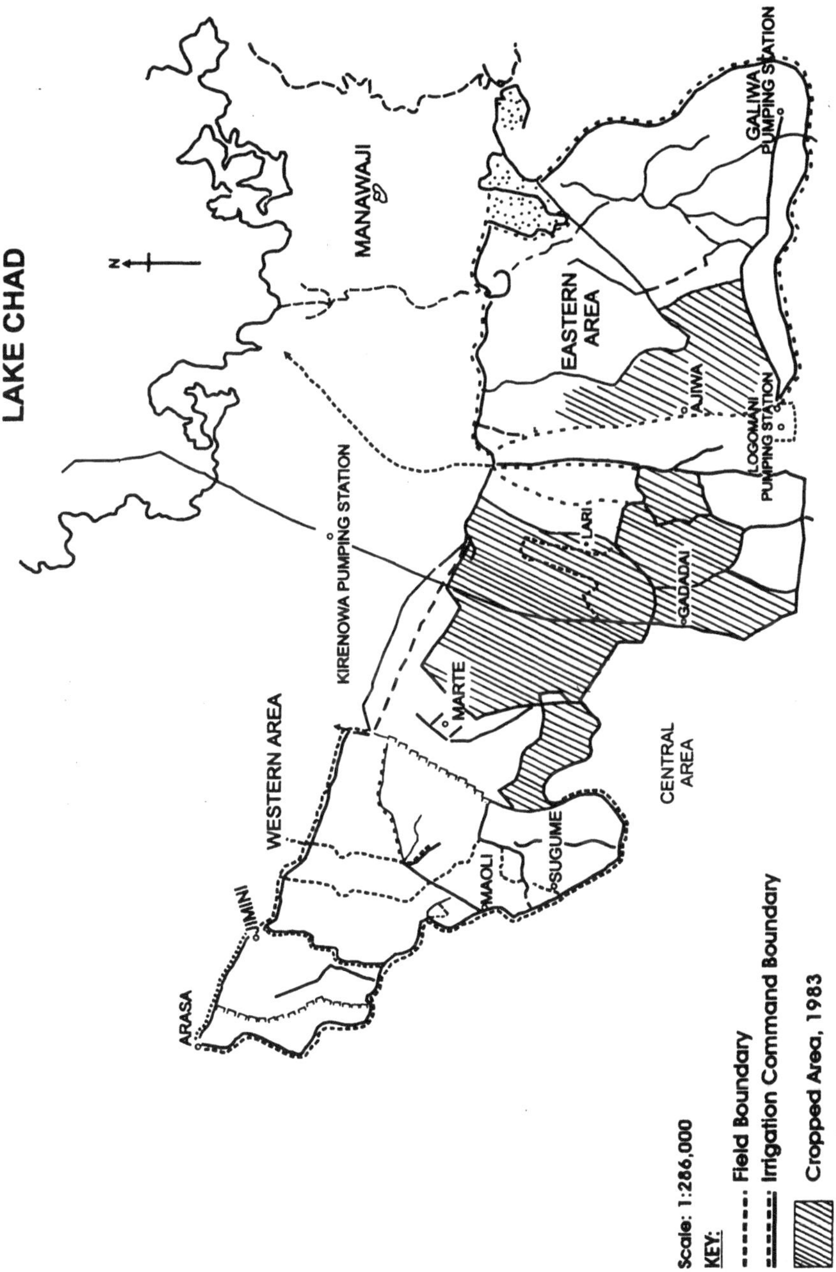

Fig. 6.4: South Chad Irrigation Project (SCIP)

cowpeas and sorghum were cultivated as well.

The insecurity of the water supply from Lake Chad due to the incessant drought in the area, has been a major problem to the irrigation project. For example, during the 1984-85 drought, the intake channel was completely exposed and, therefore, there was no water to irrigate the 10,000 ha of Stages I and II of the project. Figure 6.5 shows Lake Chad water levels at Kirenowa intake channel. The 1984 levels were completely below the critical level at which the pumping station could operate (Nwa, 1993a; Evans and Gunganesharajah, 1985; CBDA, 1984).

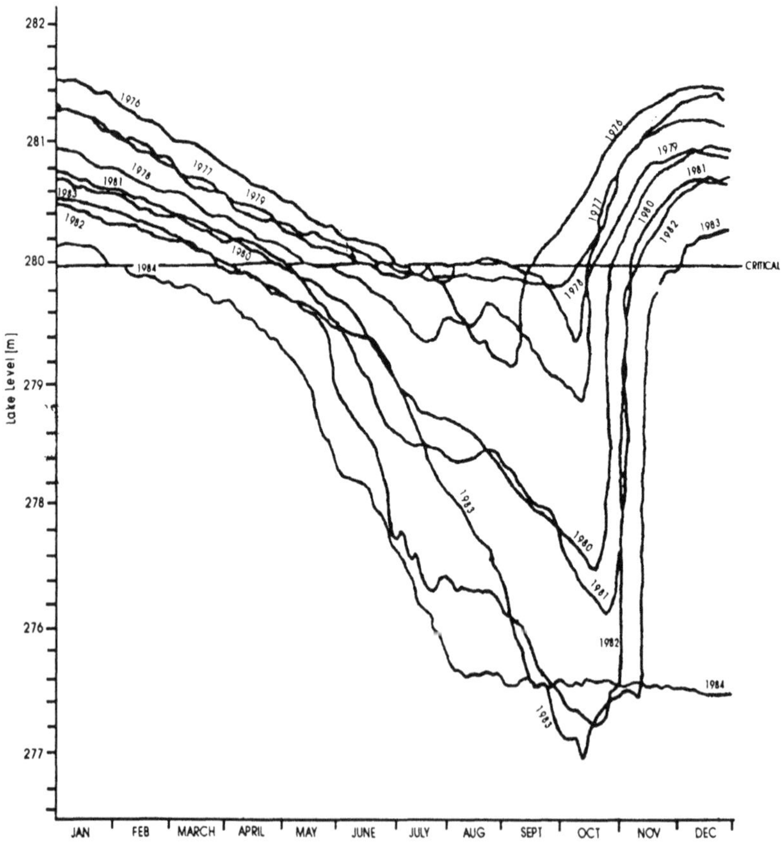

Fig. 6.5: Lake Chad Water levels and Kirenowa intake channel 1976-84

6.7.5 Bakolori Irrigation Project (BIP)

Following its study of the land and water resources of the Sokoto-Rima basin between 1963-1968, FAO recommended a plan for the development of the whole basin starting with the Bakolori Irrigation Project (BIP). The feasibility study of the project was commissioned in 1972 and the final report was submitted in 1974. Meanwhile, in 1973, the Federal Government formally established by Decree No. 33, the Sokoto-Rima River Basin Development Authority (SRRBDA) to take charge of the Bakolori Irrigation Project in particular, and water resources development and agricultural production in the basin in general (Nwa and Martins, 1982; SRRBDA, 1983). In 1975, the construction of the Bakolori dam across Sokoto river, the source of water for the project, and the construction of other irrigation infrastructure started.

BIP, covering a net area of 23,200 ha of land, is located in Talata Mafara and Maradun Local Government Areas of Zamfara State; it is approximately 110 km east of Sokoto town. Originally, 34 per cent (7,000 ha) of the project was designed for surface irrigation while 66 per cent (15,400 ha) was for sprinkler irrigation. However, due to the poor performance of the sprinkler system due to shortage of power from NEPA and poor maintenance of equipment, plans were made to convert that section of the project to the surface system in later years.

The water is conveyed from the Bakolori dam to the irrigation area through a 15 km, $30m^3/s$ capacity level supply canal; the water is further distributed to the irrigation areas through two main canals, the Right Main Canal [IR], 15 km long, and the Left Main Canal [IL], 16.6km long. Thus, the combined length of the two main canals is 31.6 km. A total of 160 pumps and 500 km of pipe lines were installed to supply irrigation water to the sprinkler irrigation area. The plan of the BIP is shown in Figure 6.6.

The first irrigation in the project took place in 1978 while the project was officially commissioned by the former President of the Federal Republic of Nigeria, Alhaji Shehu Usman Aliyu Shagari in 1983. The commissioning was at least one year behind the original completion date. This delay resulted from the disturbances caused by the villagers during 1980 and 1981. Because of the long delay in compensation and resettlement after the land expropriation, the villagers revolted, and blocked all entrances to the project area and effectively stopped the contractors from entering the area to continue with the construction work (SRRBDA, 1982, Griffith, 1983; Olofin, 1993). The Government responded by using force, moved in the mobile police, at the end of

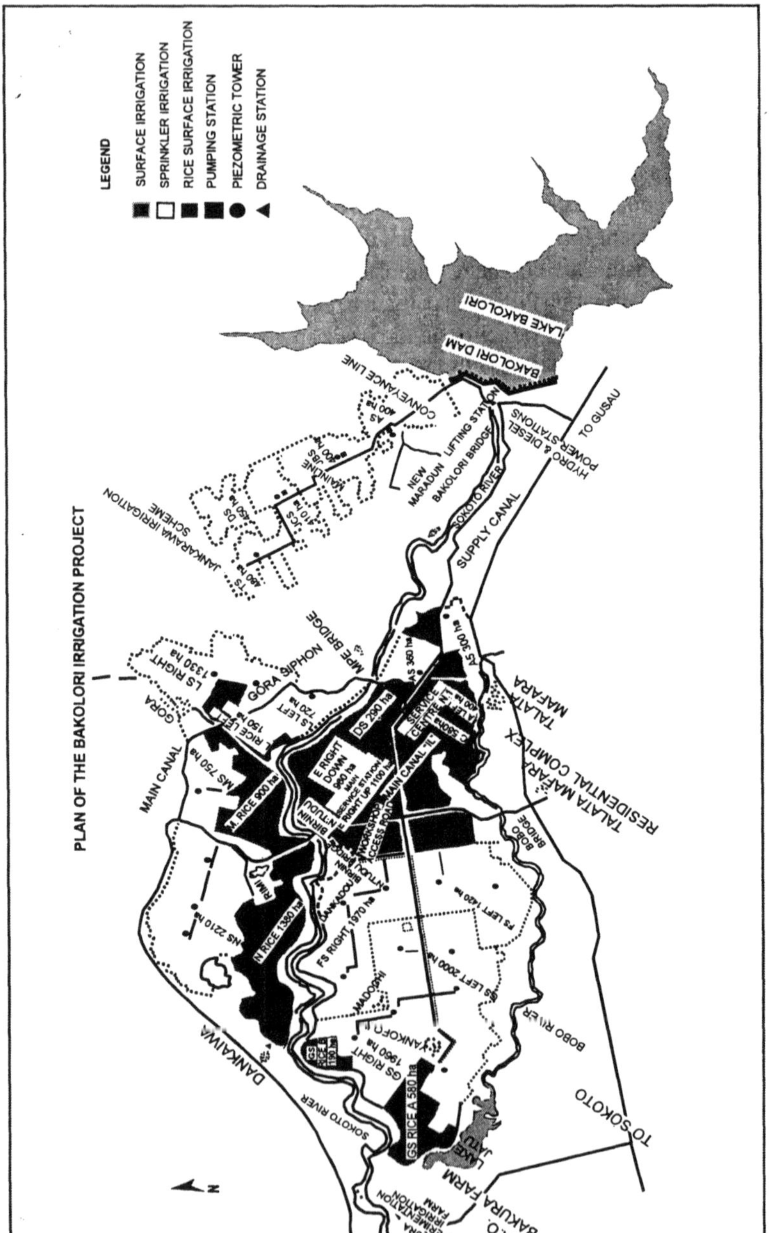

Fig. 6.6: Plan of the Bakolori Irrigation Project (BIP)

which some villages were burned and several villagers were killed. The number of people killed was variously put by the press at 14, 126, and more than three times these numbers (Griffith, 1983). Unfortunately, the true figure may never be known.

In 1992, 4,200 ha of the project were irrigated. The major irrigated crops of the scheme were wheat, maize and vegetables while rice, cotton, among other crops, were grown during the wet season with supplementary irrigation.

6.7.6 *Savannah Sugar Company Estate, Numan*

In 1970, the Federal Government of Nigeria and the Commonwealth Development Corporation (CDC) jointly established the Savannah Sugar Company Limited to investigate the possibilities of large-scale production of sugar cane in Numan Division of Adamawa State. In the same year, a small team was asked to assess the potential yields of cane and sugar, and irrigation water requirements of sugar cane in the area; the hydrology of the area was also investigated. In 1973, the CDC and others used the information collected and the experience gained to prepare a feasibility report for the project; 41,200 ha of land was identified but 13,600 ha of land was eventually proposed for the scheme.

The construction work for the scheme started in 1978 and the first irrigation in the project took place in the same year. When Kiri dam, built across Gongola river by the Federal Ministry of Industries was completed in 1982 and handed over to the Upper Benue River Basin Development Authority in 1984, arrangements were made for the Authority to supply water from the dam to the scheme. In 1992, 6000 ha of the scheme were irrigated.

6.7.7 *Peremabiri Rice Project*

In 1960, the Federal Government set up the Niger Delta Development Board and charged it with the responsibility of preparing plans for the economic and physical development of the Delta area. One of the projects identified by that Board was at Peremabiri in Diobu creek, a fresh water swamp area of the Nun river, where 120 ha of land was selected to test cropping patterns and mechanized cultivation of rice. Construction work was scheduled to start in 1965. However, nothing was done until 1967 when the feasibility study of the Peremabiri rice scheme was commissioned and a pilot project was started during the same year on 24 ha of land.

The feasibility study identified 2,500 ha of land for the proposed project. The irrigation area was to be protected from flooding by a dyke while water was to be pumped from Nun river to irrigate the rice. By Decree No. 37 of 1976, the Federal Government established the Niger Delta Basin and Rural Development Authority (NDBRDA) to take over the development of the Niger Delta from the Niger Delta Development Board whose activities were disrupted by the civil war between 1967 and 1970. The NDBRDA took over the Peremabiri rice scheme as soon as the Authority was established, and in 1979, rice was irrigated in the scheme for the first time. In 1992, 350 ha of the scheme was under irrigation.

6.7.8 *Dadin Kowa Irrigation Project*

The Gongola river basin was one of the basins in which reconnaissance studies were undertaken in 1970 to identify possible sites for irrigation projects. The studies revealed that the basin had a great potential for possible development of large-scale irrigation schemes. To give full attention to the water resources and agricultural development of the basin, the Upper Benue River Basin and Rural Development Authority was created by the Federal Government in 1976. In 1978, the Authority commissioned the feasibility study of the Dadin Kowa Irrigation Project and also started a pilot irrigation scheme of 286 ha in the area during the same year. The pilot farm was located on the right bank of Gongola river near Dadin Kowa bridge on Gombe-Biu road. The farm was irrigated by sprinkler, pumping water from Gongola river; the first irrigation took place in 1979.

The feasibility report proposed that 54,000 ha of land could be used for the Dadin Kowa Irrigation Project. The size of the project was reduced to 38,000 ha years later. The construction of the Dadin Kowa dam, the permanent source of water for the irrigation project, started in 1980 and was completed in 1988. The dam is located in Yamaltu Local Government Area of Gombe State. Up to 1992, the main irrigation scheme had not commenced, the pilot scheme area was still being operated. The main crops in the pilot scheme were rice, maize, wheat and vegetables.

6.7.9 *Lower Anambra Irrigation Project*

The Lower Anambra Irrigation Project (LAIP) is located at Omor in the Ayamelum Local Government Area of Anambra State. The feasibility study for this project started in 1976 at the end of which an area of 5,000 ha of land was found to be suitable for the irrigation project. Rice was the main crop recommended for the scheme. The construction of LAIP was started in 1981 by a Japanese Consortium supported by the Japanese International Cooperation Agency (JICA). The project was completed in 1986. The plan of LAIP is shown in Figure 6.7.

Water for the scheme is obtained from River Anambra. A pumping station at Ifite-Ogwari, lifts the water into a 16.5 km long headrace unlined main canal which conveys the water to the East and West Main Canals. The East Main Canal is 11,820 m long while the West Main Canal is 11,765 m long. The two main canals convey the water to irrigate the respective sections of the project.

The Agency which manages the scheme, the Anambra - Imo River Basin and Rural Development Authority (AIRBRDA), was created by the Federal Government in 1976. In 1985, the AIRBRDA and JICA officials carried out the first demonstration irrigation in the scheme on 35 ha of land. This demonstration which produced much higher yields than rain-fed rice (3 tonnes/ha compared to 1 tonne/ha) created a great deal of enthusiasm among the local rice farmers who had previously been undertaking only rain-fed cultivation. By 1991, the farmers in the project were already discussing the formation of an effective Water Users Association (WUA) to deal with their water supply and management problems and also a Farmers' Cooperative Society to deal with their marketing and other non-water related problems.

Two crops of rice were usually obtained in the scheme per year, one dry season crop under full irrigation, and one wet season crop under supplementary irrigation. In 1992, 3,000 ha of land in the scheme was irrigated during the dry season. Because of the nature of the soils, the current thinking is that the scheme may not be able to go beyond the 3,000 ha presently cultivated to the 5,000 ha originally proposed.

6.8 Methods of Irrigation, Irrigated Areas, Crops and Rate of Irrigation Development in Nigeria.

6.8.1 Methods of Irrigation

Two methods of irrigation dominate the Nigerian irrigation systems. These are the surface (gravity) and the sprinkler irrigation methods. Of the two methods, surface irrigation is the most important, and in this system, the furrow, basin and border methods are predominant.

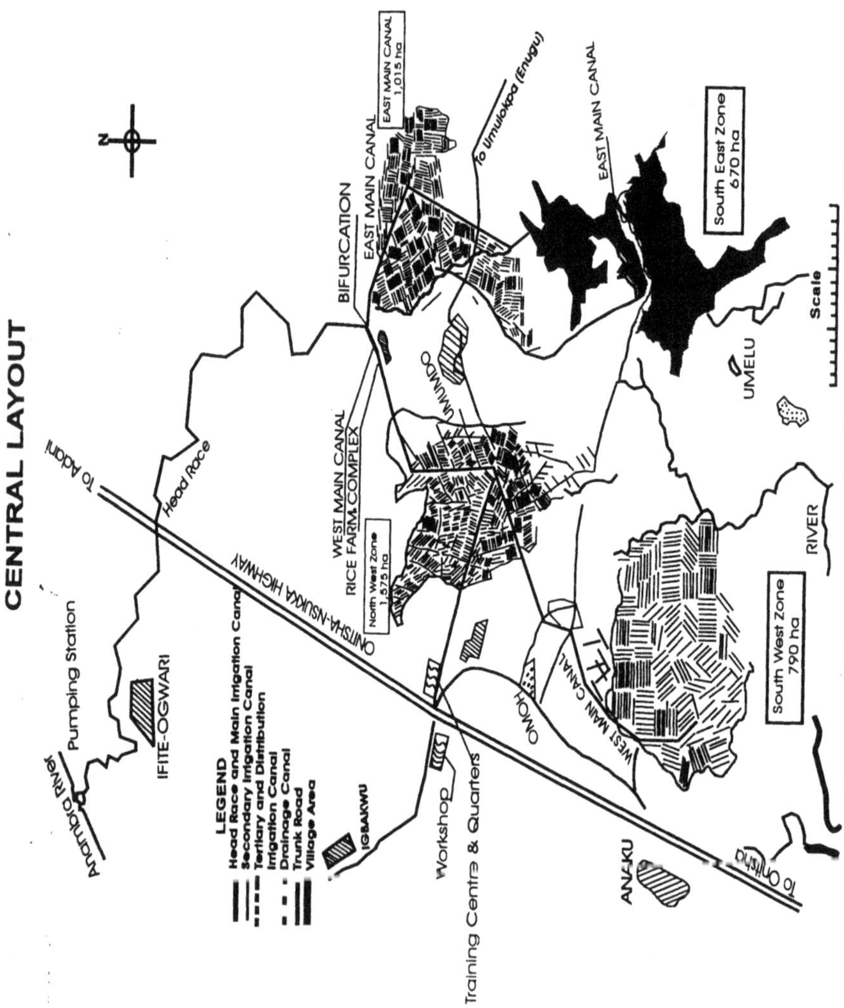

Fig. 6.7: Plan of the Lower Anambra Irrigation Project (LAIP)

Sprinkler irrigated area currently makes up an insignificant percentage of the total irrigated area in the country. The sprinkler system used is mostly the rotating head, hand-move system. Many organizations with small areas tend to start off with the sprinkler system while developing the infrastructure for large surface irrigation areas. Sokoto-Rima River Basin and Rural Development Authority initially planned to irrigate 15,400 ha by sprinkler in the Bakolori Irrigation Project. The Authority never irrigated anything near to the above figure before it started planning to convert the area to surface irrigation due to technical and management difficulties. The most successful large-scale sprinkler system to date is the irrigation project of the Nigerian Sugar Company in Bacita, Kwara State which irrigated 2,788 ha of its sugar cane estate by sprinkler in 1980. Any other irrigation systems such as the drip or trickle used in the country have been mostly experimental (Nwa, 1975). Adeniji (1992) did not uncover any other method of irrigation in his survey in 1992.

6.8.2 Irrigated Areas and Rate of Irrigation Development

Systematic data collection on water resources in general and on irrigation in particular has not yet been institutionalized in Nigeria. Even the terms used have not yet been standardized. That is why there is confusion at the moment; irrigation data now differ from source to source. The following irrigation systems are being operated in Nigeria:

1. Traditional irrigation - shadufs, calabashes and watering cans are used.
2. Lift irrigation by direct pumping - small pumps are used to pump water from rivers, streams or shallow wells for direct irrigation mostly in the fadamas.
3. Tubewell irrigation - washbore and shallow tubewell irrigation used mostly in the fadamas.
4. Diversion or flood control irrigation - water is diverted from streams or

flood water is controlled for irrigation in the fadamas.

5. Formal irrigation - the irrigated area is equipped with the necessary irrigation structures. The system is usually managed by government agencies, private companies, individual farmer or groups of farmers. The irrigated area may be large, medium or small.

Flood recession cropping, where the flooded land near rivers and lakes are cropped as the flood recedes, is not included among the irrigated areas. Quite often, the information on irrigated area in a particular year does not specify whether some areas were irrigated 2 or 3 times during the year.

Unfortunately, data on irrigated areas in Nigeria are often not given or explained in the manner indicated above. However, available data indicate that 69,000 ha were actually irrigated under the formal system, and 179,020 ha were irrigated under the combined lift, tubewell and flood control systems in 1993 (NWRMP, 1995; Mijindadi *et al*; 1993); 20,000 ha were said to be under traditional irrigation in 1985.

In the formal sector, the rate of irrigation development in Nigeria has been painfully slow. Table 6.4 emphasizes this fact. Very many reasons have been advanced for this, among which are, inadequate funding, inconsistency in government policy and manpower problems, among others. In contrast, the rate of development of fadama irrigation by the ADPs is dramatic as demonstrated in Table 6.1. This was due to the fact that the ADPs were developed as packages. Each package was made up of finance, technology, materials, equipment, extension and management, among other requirements.

Table 6.4: Extent of formal irrigation in Nigeria, 1965-1993

Year	Irrigated area ha	Rate of development ha / year	Source of information
1965	12,200	-	Rydzewski, 1972
1970	14,570	474	Elahi *et al*; 1971
1978	20,000	679	ECA, Vol, IIA, 1980
1980	30,706	5,353	Nwa and Martins, 1982
1984	36,163	1,364	Adeniji. 1985
1993	69,000	3,649	NWRMP, 1995

Sources: Nwa (1987, 1994)

6.8.3 Irrigated Crops

Various crops are irrigated in different parts of Nigeria in large, medium, small and fadama schemes. Some of the crops were irrigated using watering cans. Nwa (1994) listed the various crops grown in irrigation schemes of all sizes in 1993. The list, given in Table 6.5, was updated in 1995. The list tends to suggest that there are hardly any crops in Nigeria that cannot benefit from irrigation. The list is not given in any particular order. However, from FAO (1992) report, the top 5 crops in 1989 in Nigeria, in terms of the area irrigated, were in the order of rice, chillies and peppers, tomato, wheat and sugar cane. Table 6.6 gives the production, area and yields of the top irrigated crops in Nigeria in 1989; the area includes double cropping.

Table 6.5: List of some of the irrigated crops in Nigeria.

1. rice	7. sweet pepper	13. bitter egg-plant
2. wheat	8. sugar cane	14. lettuce
3. tomato	9. amaranthus	15. cabbage
4. maize	10. okra	16. aubergine
5. onion	11. carrot	17. celosia
6. hot pepper	12. sweet egg-plant	18. corchorus

19. solanum	33. cucumber	47. lime
20. basella	34. roselle for leaf	48. mango
21. Chinese Cabbage	35. tobacco for flower	49. guava
22. sorrel	36. cress	50. tea
23. sweet potato	37. spinach	51. pineapple
24. lamsir	38. cucurbits	52. cashew
25. garlic	39. beetroot	53. banana
26. cowpea	40. radish	54. plantain
27. french bean	41. Swiss chard	55. lemon
28. water mellon	42. parsnip	56. grapefruit
29. water leaf	43 turnip	57. tangerine
30. pumpkin	44. cassava	58. tangelo
31. peas	45. pawpaw	59. coconut
32. squash	46. orange	60. oil palm

Source: Nwa, 1994a.

Table 6.6: Irrigated crops, area, yields and production, 1989

Crop	Area ha	Avg. yield tonnes/ha	Production tonnes
Rice	714,000	2.3	1,642,000
Chillies and peppers	80,000	10.0	800.000
Tomato	52,000	12.2	650,000
Wheat	50,000	1.2	60,000
Sugar cane	26,000	52.0	1,350,000
Green maize	20,000	7.5	150,000
Onion	5,000	40.0	200,000
Polato	3,000	14.0	42.000

Source: FAO, 1992

Chapter 7

Evolution of the Federal Ministry of Water Resources and Irrigation Administration in Nigeria

7.1 Irrigation Administration During the Colonial Period, from 1900 to 1960

The administration of irrigation has always been very closely tied up with the administration of agriculture in Nigeria. Therefore, the starting point of agricultural administration will, inevitably, provide the starting point for irrigation administration in the country. Probably 1893 could be taken as the beginning of agricultural administration in Nigeria when Sir Claude Macdonald set up a botanical garden and research station in Lagos. In 1899, the British Cotton Growing Association (BCGA) had set up an agricultural experiment station at Moor Plantation near Ibadan but later in the same year, the Colonial Government acquired the 252 ha piece of land for its own agricultural purposes (Green Revolution National Committee, 1981).

Between 1906-9 there was in existence the post of "Superintendent of Agriculture for British West African Colonies and Protectorate"; the officer who occupied this position travelled between the territories concerned and was able to report directly to the Colonial Office in London. Although the post of Director of Agriculture in Southern Nigeria existed in 1909, the position was not filled. It was in 1912 that separate departments of agriculture were set up for Northern and Southern Nigeria with Samaru, near Zaria, as the headquarters for Northern Nigeria and Moor Plantation as headquarters for Southern Nigeria. Following the amalgamation of Northern and Southern Provinces in 1914, the Northern and Southern Departments of Agriculture were amalgamated in 1921 to form a new Department of Agriculture for Nigeria with its headquarters at Moor Plantation; Mr. O.T. Faulkner was its first director (Green Revolution National Committee, 1981). This arrangement did not mean that the regional departments of agriculture were abolished.

The first irrigation engineer in Nigeria, Colonel M.R. Collins, was appointed in 1925 and sent to Sokoto to investigate "the possibility of water control for rice growing in the valley of the Sokoto river" (KMAG,

File No. 11019, 1919); it was Colonel Collins who started the construction of the Kware irrigation scheme as reported in Chapter 4. The second irrigation engineer in Nigeria, Mr. A.L.V. Long was sent to Sokoto in 1926 to replace Colonel Collins. It was in 1949, more than 20 years after Mr. Long, that the third irrigation engineer in Nigeria, Mr. John Kitching, was sent to Sokoto. It was in that year (1949) that the first Irrigation Division in Nigeria was established in the Agriculture Department of Northern Nigeria. Although the headquarters of the Department of Agriculture of the North was in Samaru, the headquarters of the Irrigation Division was in Sokoto.

Probably to further strengthen the unification process, the Central Department of Agriculture appointed, in 1950, three Deputy Directors of Agriculture each to be responsible for the Northern, Eastern and Western Provinces which had been created in 1939. When the Macpherson Constitution became effective in 1951, the regional administration in each of the three provinces was upgraded to a regional government; the Director of the Central Department of Agriculture became known as the Inspector-General of Agriculture. Mr. O.T. Faulkner was the Central Director of Agriculture from 1921-36 and was succeeded by Captain J.R. Mackle from 1936-45; Mr. A. G. Beattie who succeeded Captain Mackle was Director of Agriculture from 1945-50 and then became the Inspector-General of Agriculture in 1951 until he retired at the beginning of 1953. Meanwhile in 1952, the National Department of Agriculture became part of the Central Ministry of Lands and Natural Resources under a Federal Minister, Honourable K.O. Mbadiwe.

When the first Federal Constitution came into force in 1954 (Crowder, 1966), the regional governments were given the sole responsibility for agricultural and water development, including irrigation development. In 1956, Mr. John Kitching was appointed the Assistant Director of Agriculture in charge of irrigation in Northern Region. In the same year, an irrigation station was opened in Maigana near Zaria in addition to those already opened in Kano in 1955, Maiduguri in 1954 and in Bida and Pateggi before 1954. When a regional self-government was granted to the Eastern and Western Regions in 1957, another constitution was introduced into Nigeria.

Because of these political changes, the Department of Agriculture of Northern Region became Ministry of Agriculture, in line with those in the East and West, in 1957 and moved from Samaru to Kaduna. Samaru became the headquarters of the Research and Specialist Division of the Ministry and the Irrigation Division moved from Sokoto to Kaduna. In 1959, a Hydrological Section was set up in the Irrigation Division of the Ministry of Agriculture of Northern Region to develop a network of hydrological stations and organize in-field training services (FAO, 1966). During this period, very little or no discussion on irrigation took place in the Eastern and Western Regions.

7.2 Irrigation Administration, 1961-74

Following the attainment of political independence in 1960, the Federal Ministry of Agriculture and Natural Resources (FMANR) was created in 1965. However, its role was formally confined to the administration of agricultural research. Mr. G.A. Ige was the first Permanent Secretary of this new ministry from 1965-68 and was succeeded by Dr. Bukar Shaib from 1968-75. In 1966, the responsibility for water resources was officially given to FMANR. Before this time, precisely in 1960, the Nigerian Hydrological Technical Committee (NHTC) had been set up to advise government on matters of water resources and water resources data collection. However, a sub-committee of NHTC was to lament in 1974 that the sub-committee "is aware of the advisory role of the Hydrological Technical Committee, but views with great concern the fact that within the last 10 years the recommendations and resolutions that have been submitted by the Hydrological Technical Committee have neither been implemented nor even considered, and that these now lie in closed files" (NHTC, 1974). The Northern Region created a Ministry of Water Resources in 1965.

Resulting from the recommendation of FAO (1966) that FMANR should be expanded to include a number of technical and professional departments, the Federal Department of Agriculture (FDA) was created in 1970 with Mr. B. S. Oloruntoba as its first Director. Land and Water Resources Division was one of the divisions created for FDA and it was to coordinate the activities and provide technical assistance to the states in the field of land tenure, soil and water conservation, land capability surveys, water resources survey and irrigation development. The headquarters of the division was in Kaduna. In 1973, Mr. N.O. Popoola was appointed to head this division as the first Chief Land and Water Resources Officer.

There was very little change in irrigation administration in the Eastern and Western Regions during this period. In the first National Development Plan document (1962-68), while the North already had an established structure for irrigation administration, the East and the West did not even mention irrigation at all in the document. As indicated in Section 6.4, the first irrigation engineer in the East was recruited in 1963 and was attached to the Division of Agriculture and was responsible to the division's Adviser on rice promotion.

In 1963, the Mid-Western Region was created from the Western Region and, therefore, Nigeria had the four regions of North, East, West and Mid-West, and in 1967 these four regions were broken into 12 states by the Military Head of State, Lt. Col. Yakubu Gowon. Because of the state creation, all the regional ministries of agriculture were dissolved and in 1968 each state created its Ministry of Natural Resources (MNR). The states in the North carried the Irrigation Division from the Northern Region into their MNR. It was in the late 1960s and early 1970s that irrigation units and divisions were created in MNR of the eastern, western and mid-western states.

7.3 Irrigation Administration, 1975-1999

In early 1975, the Federal Department of Water Resources (FDWR) was created in the FMANR by merging the Land and Water Resources Division of the FDA and the Water Supply and Engineering Division of the Geological Surveys Department. Mr. N.O. Popoola became the first Director of the FDWR under the FMANR. The Department had three technical divisions of Irrigation and Water Supply, Hydrology and Hydrogeology, and Planning and Design. In April 1975, the Federal Ministry of Water Resources (FMWR) was created (Nwa, 1987) with Professor I.U.W. Osisiogu as the Federal Commissioner for Water Resources (1975-1977), Dr. Bukar Shaib as Permanent Secretary (1975-1978), and Mr. N.O. Popoola as Director, FDWR (1975-1979). The ministry had only one technical department, the FDWR, with the main function of coordinating the technical activities of both the Federal and State agencies in the field of water resources.

At the time the FMWR was created in 1975, the National Council for Water Resources (NCWR) and the National Technical Committee on Water Resources (NTCWR) were also established. The Federal Minister of Water Resources was made the Chairman of the NCWR and the membership included all the commissioners responsible for water resources in the states and the Chairman of the Federal Capital Territory Water Resources Agency. The functions of the Council included the coordination of the policies and activities of the various state governments in water resources exploitation and development. The inaugural meeting of the Council took place in 1981. The Council receives technical advice from NTCWR which is chaired by the Director-General or Permanent Secretary of Water Resources and whose membership includes representatives of all the organizations involved in water resources development and exploitation including the universities, polytechnics and private consultants. NTCWR does its work through the sub-committees on:

1. Irrigation and Drainage;
2. Hydrology and Hydrogeology;
3. Dams and Reservoir Operations;
4. Water Supply and Quality Control;
5. Manpower Development; and
6. Rural Development.

The above sub-committees did not start functioning immediately they were created. At the 25th session of the Nigerian Hydrological Technical Committee (NHTC) held at Kaduna in 1977, six Working Groups were established. One of such groups was the Working Group on Irrigation and Drainage with Mr. A.O. Omolokun as the Convenor and Dr. E.U. Nwa as the Secretary. When the National Technical Committee on Water Resources (NTCWR) was inaugurated in 1981, also in Kaduna, to take over the functions of NHTC, the National Sub-Committee on Irrigation and Drainage (NSCID) was also inaugurated to take over the functions of the Working Group on Irrigation and Drainage of NHTC. Alhaji Ahmadu Rufa'i was the Chairman of NSCID (1981-1986) while Professor E.U. Nwa was Vice-Chairman (1981-1986); Professor Nwa took over the chairmanship from 1986-1995. The function of NSCID was to advise NTCWR on all matters relating to irrigation and drainage.

It was already pointed out in Section 6.5.4 that the Federal Department of Rural Development (FDRD) was created in 1976 within the FMANR and, therefore, the ministry became the Federal Ministry of Agriculture and Rural Development (FMARD). Thus, when Water Resources was first merged with Agriculture in 1977 by the Head of State, General Olusegun Obasanjo, FMWR was merged with the FMARD to form the Federal Ministry of Agriculture, Water Resources and Rural Development (FMAWRRD). Mr. B.O.W. Mafeni was the Federal Commissioner for the new ministry between 1977-1979.

When Alhaji Shehu Aliyu Shagari was sworn in as a civilian President on 1st October, 1979, he re-created the FMWR with Mr. Ndagi

Mamudu as the Minister for Water Resources (1979-81), Dr. E. Yahaya Atanu, Minister of State for Water Resources (1979-81), Mr. M.K.O. Williams, Permanent Secretary (1979-83), and Dr. B. Martins as Director, FDWR (1979-84). In 1981, Dr. E. Yahaya Atanu became the Minister (1981-83) and Chief Eteng Okoi-Obuli became the Minister of State for Water Resources (1981-83); and in 1983, Alhaji Ismaila Isa became the Minister of Water Resources (Oct 1983-Dec 1983), Mr. N.O. Popoola became the Coordinating Director (Oct 1979-Mar 1984).

General Muhammadu Buhari took over power as the Head of State in December, 1983 and in January, 1984 he re-merged the FMWR with the FMARD to form the FMAWRRD. Dr. Bukar Shaib was appointed the Minister for the new ministry (Jan. 1984-Sept. 1985), Mr. N.O. Popoola became the Permanent Secretary (March 1984-Jan. 1986), and Dr. Umar Kawu became the Director, FDWR (Mar 1984-Dec 1989). General Ibrahim Badamosi Babangida displaced General Buhari in August, 1985 and became the first Military President of Nigeria. In December, 1989, President Babangida re-created the FMWR and in January 1990, Alhaji Bunu Sheriff Musa was appointed the Minister of Water Resources (Jan 1990-Aug 1990) and Dr. Alex Usman Kadiri was appointed the Director-General (Jan 1990-Feb 1994); in August 1990, Alhaji Abubakar Habu Hashidu took over as Minister of Water Resources (Aug 1990-Jan 1992).

When the FMWR ceased to be one-Department Ministry in January 1990, the Federal Department of Irrigation and Drainage (FDID) was created for the first time; Engr. H. Hamisu Ibrahim became the first Director of the Department (Jan 1990-Oct 1993). Engr. Ibrahim was succeeded by Engr. Inuwa K. Musa as Director in October, 1993. The creation of departments in the FMWR in 1990 did not end the "permanent state of reorganization" of the ministry according to Nwa (1994c). In 1992, the ministry was re-merged with the FMARD by President Babangida to form the FMAWRRD with Alhaji Abubakar Habu Hashidu as the Minister. In January, 1993, Alhaji (Dr.) Garba Ja Abdulkadri was appointed the Secretary of Agriculture, Water Resources and Rural Development while Alhaji Isah Mohammed was appointed Secretary of State for Water Resources by President Babangida. However, in August, 1993 when Chief Ernest Adegunle Oladeinde Shonekan became the Head of the Interim National Government, FMWR was re-created and Rural Development was added to it to form the Federal Ministry of Water Resources and Rural Development (FMWRRD) and Alhaji Isah Mohammed became the Secretary of Water Resources and Rural Development.

General Sani Abacha took over as Head of State in November, 1993 and in February, 1994, Alhaji Isah Mohammed was reappointed as the Minister of Water Resources and Rural Development. Mr. J.A. Hanwood was the Acting Director-General (Feb 1994-May 1994) of the Ministry until Engr. A. Aliyu Dogondaji took over as the Director-General in May, 1994. Alhaji Aliyu Yelwa replaced Alhaji Isah Mohammed as the Minister of Water Resources and Rural Development in March, 1995. Alhaji Hamza Sakwa replaced Alhaji A Yelwa as Minister in 1997 and Engr. A. Aliyu Dogondaji was replaced by Mr Wada M. Aliyu as the Permanent Secretary in 1998. On taking office as the Civillian President on 29th May, 1999, General Olusegun Obasanjo transferred Rural Development back to the Ministry of Agriculture and appointed Col. Bello Kaliel as the Minister of Water Resources.

As indicated in Chapter 8, two River Basin Development Authorities (RBDAs) were created in 1973; the number was increased to 11 in 1976 to cover the entire country. The RBDAs were increased to 18 in 1984 and Rural Development was added to their functions, then reduced back to 11 in 1986 and Rural Development was transferred to DFRRI; the current number of 12 has been maintained since 1994. The RBDAs serve as the execution arm of the FMWRRD; they are responsible for managing most of the medium and large-scale irrigation schemes in the country.

The Agricultural Development Projects (ADPs) were launched by the Federal Government in 1975. However, it was between 1981 and 1984 that the ADPs got involved in small scale irrigation development. This involvement was greatly accelerated in 1992 when the Federal Government obtained a World Bank loan to launch the National Fadama Development Project (NFDP). The use of small pumps and tubewells for irrigation was accelerated during this period. The ADPs are supervised by the Federal Agricultural Coordinating Unit (FACU) on behalf of the Federal Ministry of Agriculture and Natural Resources.

Irrigation in the states is managed by the State Ministry of Agriculture and Natural Resources or the State Ministry of Water Resources and Rural Development. At the state level, the irrigation projects are largely small-scale.

To a smaller extent at the moment, commercial companies and small independent farmers are also involved in irrigation development and management. The total hecterage controlled currently by these groups is small but this sector has a high potential for growth in the future.

7.4 Current Structure for Irrigation Development and Management

Irrigation development and management in Nigeria are currently undertaken at four levels, namely; Federal, State, Local Government and Private Sector levels. These arrangements are summarized below:

a. Federal Level
   1. National Council for Water Resources and Rural Development (NCWRRD)
      - National Technical Committee on Water Resources and Rural Development (NTCWRRD)
   2. Federal Ministry of Water Resources and Rural Development (FMWRRD)
      - Department of Irrigation and Drainage
      - RBDAs
      - National Water Resources Institute
   3. Federal Ministry of Agriculture and Natural Resources (FMANR)
      - FACU - to supervise the ADPs
      - National Agricultural Research Institutes — to organize research.

b. State Government Level
   - State Ministry of Agriculture and Natural Resources or State Ministry of Water Resources and Rural Development.
   - ADPs

c. Local Government Level
   - Local Government Agricultural/Irrigation Units

d. Private Sector
   - Commercial Companies
   - Small Independent Farmers

The organogram of the FMWRRD, Federal Department of Irrigation and Drainage, one of the RBDAs and FMANR are shown in Figures 7.1 - 7.4.

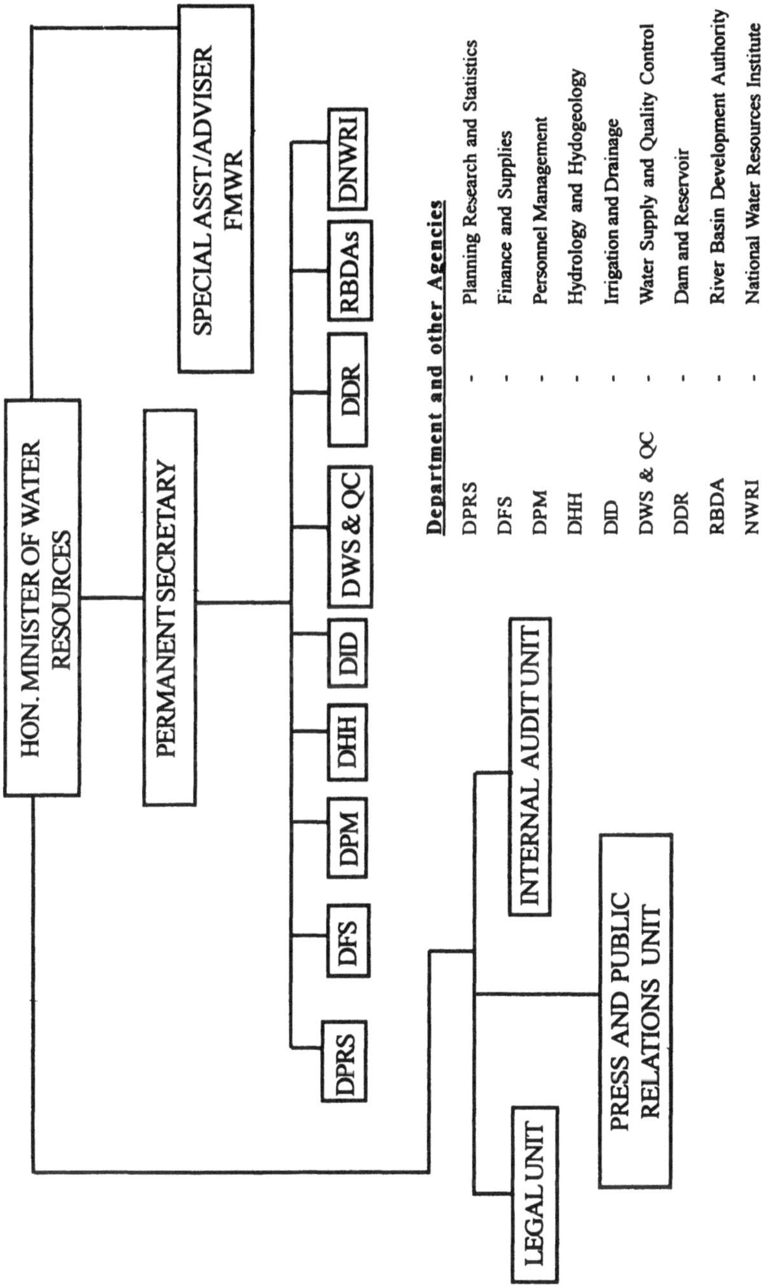

Fig. 7.1. Organizational Chart of the Federal Ministry of Water Resources

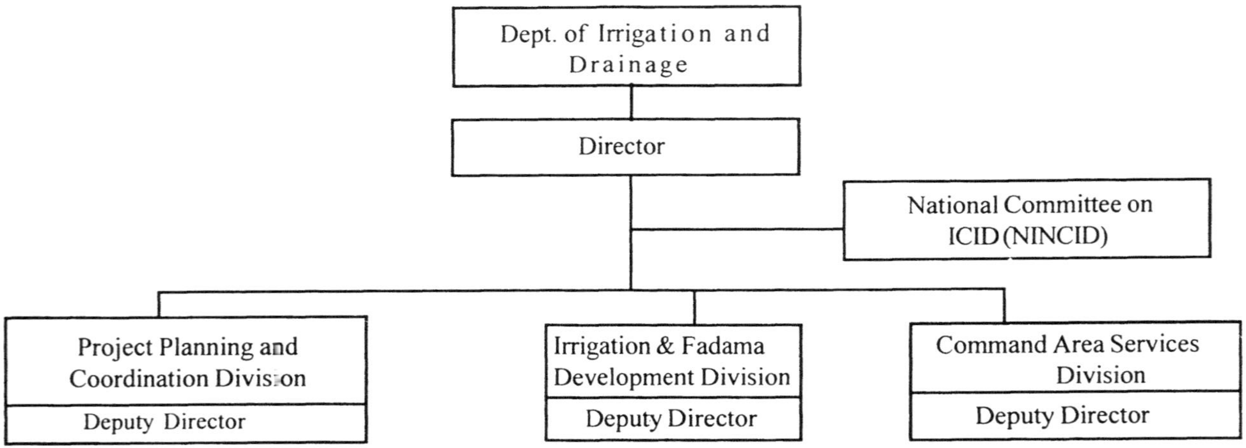

* Policy and Planning
* Environmental Impact Assessment
* Monitoring & Evaluation
* Training, Workshops & Seminars

* Fadama Project Planning Formula & Coordination
* Engineering Design & Construction
* Operation & Maintenance
* Water Users & Fadama Users Association

* Technical Services
* Extension Services
* Adaptive Research
* On-farm Investigations & Surveys

Fig.7.2: Organogram of the Federal Department of Irrigation and Drainage

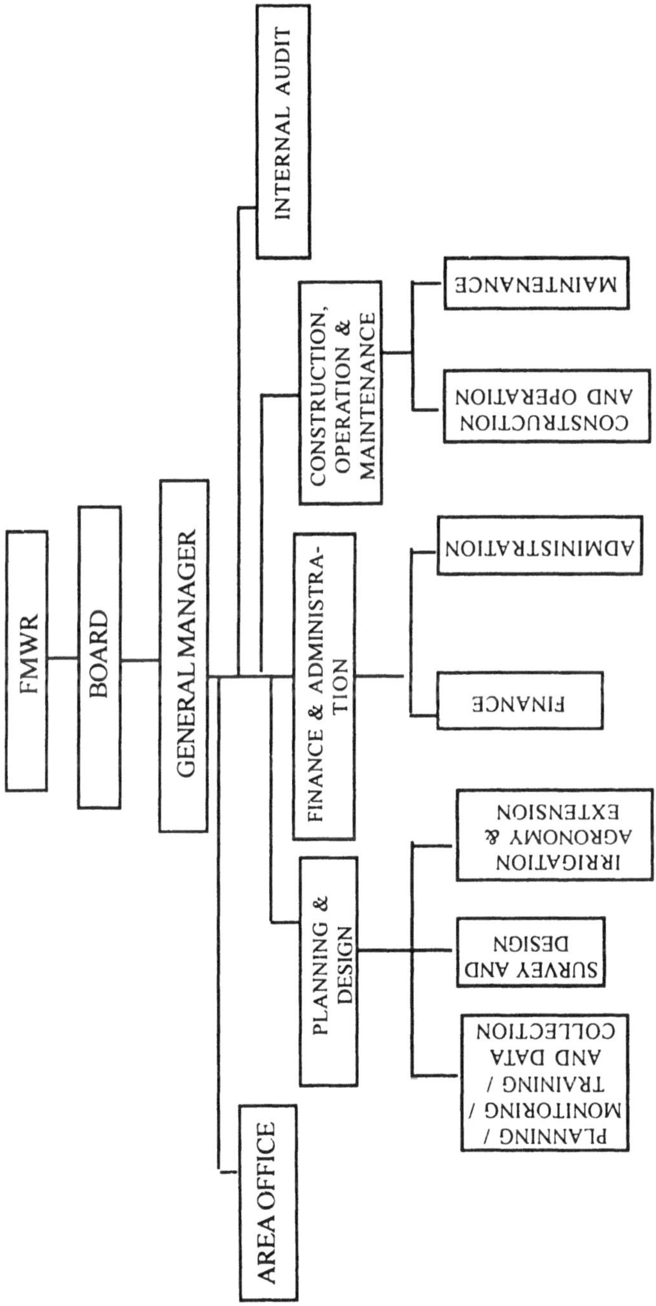

Fig. 7.3: Organogram of a River Basin Development Authority (Group A)

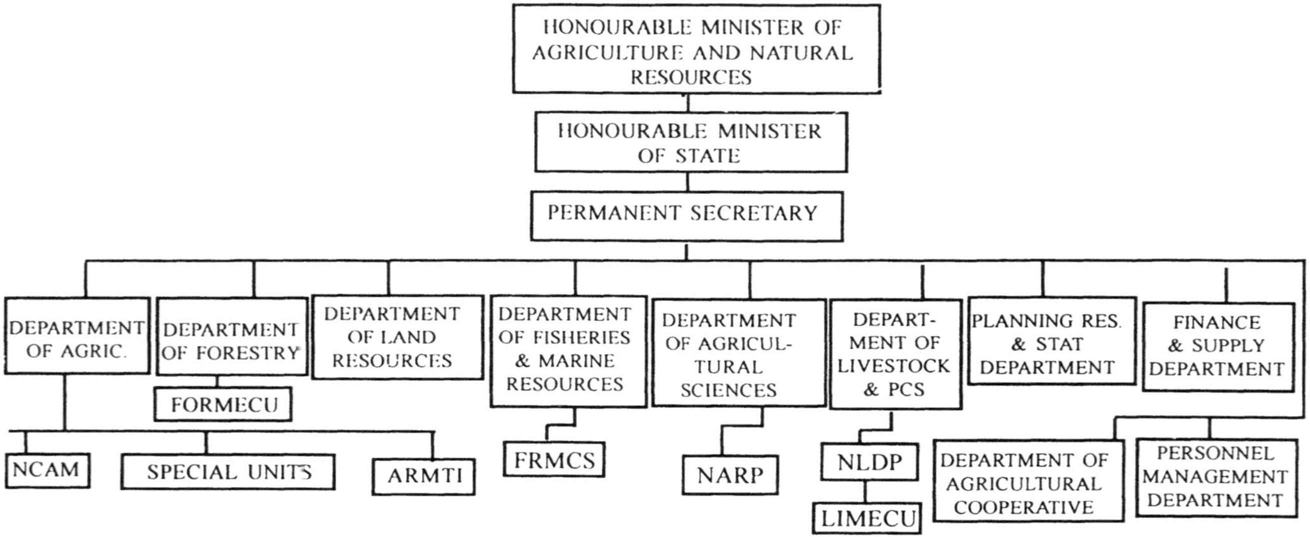

1.	ARMTI	=	Agricultural and Rural Management Training Institute	2.	FORMECU	=	Foresty Monitoring, Evaluation and Coordinating Unit
3.	FRMCS	=	Fisheries Research Monitoring, Control and Surveillance	4.	LIMECU	=	Livestock Monitoring, Evaluation and Coordinating Unit
5.	NARP	=	National Agricultural Research Project	6.	NCAM	=	National Centre for Agricultural Mechanization
7.	NLDP	=	National Livestock Development Project	8.	PCS	=	Pest Control Services

Fig. 7.4: Organogram of the Federal Ministry of Agriculture and Natural Resources

7.5 Legal Instruments for Water Resources and Irrigation Management.

The management of rivers and lakes which cross international and state boundaries is Federal responsibility under the constitution.

7.5.1 Water Resources Decree No 101 of 1993.

According to the decree:

> the right in the use and control of all surface and groundwater and of all water in water - course affecting more than one State as described in Schedule to this Decree, together with the bed and banks therefore by virtue of this Decree and without further assurance vested in the Government of the Federation for the purpose of:
> a) promoting the optimum planning, development and the use of Nigeria's water resources.
> b) ensuring the coordination of such activities as are likely to influence the quality, quantity, distribution, use and management of water;
> c) ensuring the application of appropriate standards and techniques for the investigation, use, control, protection, management and administration of water resources, *among others.*

In the case of the control of groundwater, the decree states, among several other things, that:

> Notwithstanding the provisions of section 2 of this Decree, the Secretary shall have power to:
> a) define the places from which or the manner in which and the times at which such water may be taken or used;
> b) fix, in times of actual or anticipated shortage of water, amounts which may be taken by any person for such purposes;
> c) prohibit temporarily or permanently, the taking or use of water for any such purposes, when in his opinion the taking of such water would be hazardous to health;
> d) regulate the place, depth, manners of construction or mode of operation of any borehole or well.

7.5.2 Rules and Regulations for the Administration and Enforcement of the Provision of the Water Resources Decree No. 101 of 1993.

The objective of the above document is to set out the rules and regulations approved by the government for the administration and enforcement of Decree No. 101 of 1993. For example, the document specifies the levels of substances such as nitrate and sulfates that could be allowed into surface and ground water.

7.5.3 River Basin Development Authorities Decree No. 35 of 1987.

In addition to establishing the River Basin Development Authorities (RBRDAs), Decree No 35 of 1987 sets out the functions of the RBRDAs; these are discussed in section 8.4.

7.5.4 Federal Environmental Protection Agency Decree No. 59 of 1992.

The Federal Environmental Protection Agency (FEPA) was established and given the responsibility "for the protection and development of the environment and biodiversity conservation and sustainable development of Nigeria's natural resources ..."

7.5.5 Environmental Impact Assessment Decree No. 86 of 1992.

The Environmental Impact Assessment Decree was intended to give FEPA the authority to ensure that the environmental impact of any activity are first of all, taken into account before the activity is undertaken. All the above decrees are the legal instruments that may be used to establish rules and regulations, guidelines, procedures and by-laws to ensure that irrigation and other water resources systems are exploited and managed in a sustainable manner.

Chapter 8

The Concept and Evolution of River Basin Development

8.1 Background

The River Basin Development concept is not new. The ancient developments in the Nile valley and in the valleys of the Tigris and Euphrates are cases in points. However, the Tennessee Valley Authority (TVA) initiated in the United States of America in 1933 may be considered as the first of the modern River Basin Development Authorities (Huxley, 1945). Since then a number of river valley authorities patterned after TVA have been set up in different parts of the world. One of such is the Damodar Valley Corporation, "India's experiment with the TVA model" (Hamilton, 1969).

8.2 Evolution of River Basin Development in Nigeria

In 1960, the first River Basin and Rural Development Authority in Nigeria was set up by the Federal Government. It was called the Niger Delta Development Board. It was charged with the responsibility of preparing plans for the economic and physical development of the Delta area. As its contribution to the water resources development of the area, the Board embarked on flood protection schemes and made a start at the collection of hydrological data (Nwa and Martins, 1982).

In 1963, as indicated in Section 6.2, the Food and Agriculture Organization of the United Nations, embarked on the study of the land and water resources of the Sokoto-Rima basin, and in 1965, FAO and UNESCO conducted a "study of the water resources in the Chad basin." Also in 1965, the Bureau of Reclamation of the United States Department of Interior, carried out a study of the Kano River Basin which led to the establishment of the Kano River Project in 1968 by the Kano State Government. As a result of these studies and activities in these basins and the severe Sahelian drought of 1972-73 the Federal Government formally established two river basin authorities, the Chad Basin Development Authority by Decree No. 32, and the Sokoto-Rima Basin Development Authority by Decree No. 33, both of 14th August, 1973. These authorities were to coordinate water resources and irrigation development in their respective catchments.

In 1976, the Federal Government divided the whole country into 11 "River Basins." Ten of these basins were established by Decree No. 25 of 15th June, 1976 while the 11th, the Niger Delta Basin Development Authority, was established by Decree No. 37 of 3rd August, 1976. The continued existence of these RBDAs was reconfirmed by Decree No. 87 of 28th September, 1979.

8.3 Expansion and Contraction of the RBDAs

In 1984, the number of the RBDAs was increased from 11 to 18 and Rural Development added to their functions. The 18 basin authorities and their headquarters were the following:

1. Anambra River Basin and Rural Development Authority - Enugu
2. Benin River Basin and Rural Development Authority - Benin City
3. Chad Basin and Rural Development Authority - Maiduguri
4. Cross River Basin and Rural Development Authority - Calabar
5. Gongola-Jama'are River Basin and Rural Development Authority - Dadin Kowa
6. Imo River Basin and Rural Development Authority - Owerri
7. Kaduna-Karaduwa River Basin and Rural Development Authority - Kaduna
8. Kano-Hadejia River Basin and Rural Development Authority - Kano
9. Lower Benue River Basin and Rural Development Authority - Makurdi
10. Lower Niger River Basin and Rural Development Authority - Ilorin
11. Lower Ogun-Oshun River Basin and Rural Development Authority - Abeokuta
12. Mada-Dep River Basin and Rural Development Authority - Lafia
13. Niger Delta Basin and Rural Development Authority - Port Harcourt
14. Owena River Basin and Rural Development Authority - Akure

15. Sokoto-Rima River Basin and Rural Development Authority - Sokoto
16. Upper Benue River Basin and Rural Development Authority - Yola
17. Upper Niger River Basin and Rural Development Authority - Minna
18. Upper Ogun-Oshun River Basin and Rural Development Authority - Ibadan

The RBDAs were reduced from 18 back to 11 in 1986 and Rural Development was transferred to DFFRI in 1993, the Niger River Basin was split into two, the Lower - and the Upper Niger River Basin Development Authorities to bring the number of the authorities to the current level of 12. The 12 basins and their headquarters are as follows:

1. Anambra-Imo River Basin Development Authority - Owerri
2. Benin-Owena River Basin Development Authority - Benin City
3. Chad Basin Rural Development Authority - Maiduguri
4. Cross River Basin Development Authority - Calabar
5. Hadejia-Jama'are River Basin Development Authority - Kano
6. Lower Benue River Basin Development Authority - Makurdi
7. Lower Niger River Basin Development Authority - Ilorin
8. Niger Delta Basin Development Authority - Port Harcourt
9. Ogun-Oshun River Basin Rural Development Authority - Abeokuta
10. Sokoto-Rima River Basin Development Authority - Sokoto
11. Upper Benue River Basin Development Authority - Yola
12. Upper Niger River Basin Development Authority - Minna

The 12 RBDAs are shown in the map of Figure 8.1.

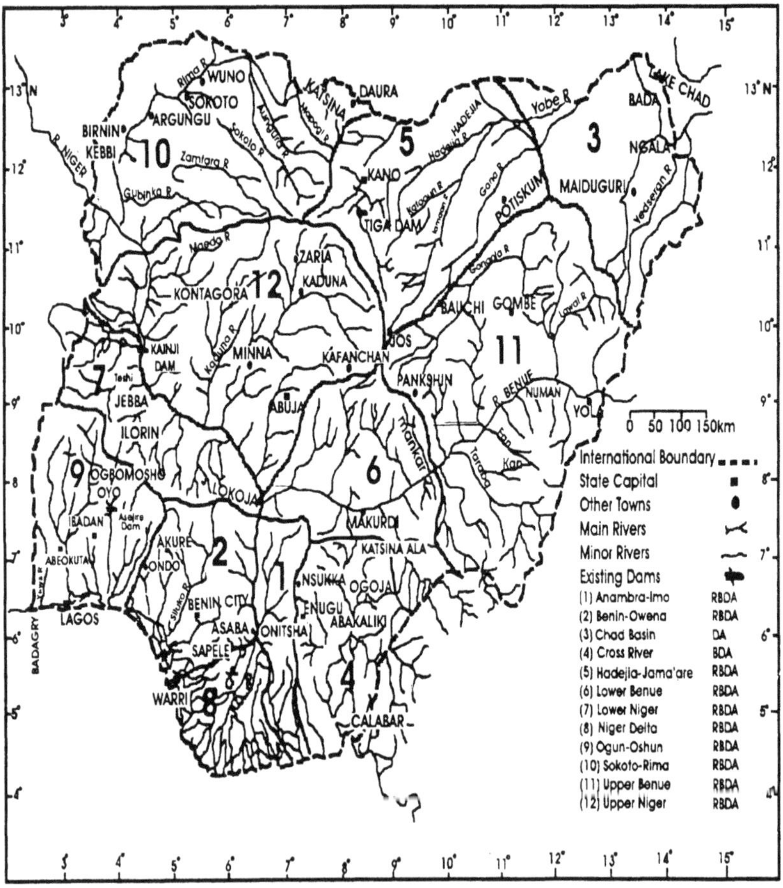

Fig. 8.1: Surface water sources and the River Basin Development Authorities of Nigeria

8.4 Decree No. 35 of 1987

Decree No, 87 of 1979 that reconfirmed the existence of the RBDAs also gave the authorities a very wide mandate including direct agricultural production. The wide mandate of the RBRDAs was reviewed by Decree No. 35 of 1987 and the authorities were divested of their direct agricultural production functions. According to the Decree (Fed. Rep. of Nigeria, 1987), the functions of each authority shall be:

a. to undertake comprehensive development of both surface and underground water resources for multipurpose use with particular emphasis on the provision of irrigation infrastructure and the control of floods and erosion and for watershed management;
b. to construct, operate and maintain dams, dykes, polders, wells, boreholes, irrigation and drainage systems, and other works necessary for the achievement of the Authority's functions and hand over all lands to be cultivated under the irrigation scheme to the farmers;
c. to supply water from the Authority's completed storage schemes to all users for a fee to be determined by the Authority concerned, with the approval of the Minister;
d. to construct, operate and maintain infrastructural services such as roads and bridges linking project sites; provided that such infrastructural services are included and form an integral part of the list of approved projects;
e. to develop and keep up-to-date comprehensive water resources masterplan, identifying all water resources requirements in the Authority's area of operation, through adequate collection and collation of water resources, water use, socio-economic and environmental data of the River Basin.

8.5 Partial Commercialisation of the RBDAs

The Technical Committee on Privatization and Commercialization (TCPC) was established by the Federal Government in July, 1988 and in 1989, TCPC introduced the concept of partial commercialization for the RBDAs to make them more profitable and less dependent on the national treasury (Zayyad, 1992). Based on their level of development and the size and complexity of their operations, TCPC developed new organizational and management structures for the authorities and categorized them into three groups. The groups and their levels of Government subvention were as follows:

Group A: This group was made up of three Authorities, namely;

- Chad Basin
- Hadejia-Jama'are, and
- Sokoto-Rima.

Fifty per cent of their recurrent subvention was to be withdrawn in 1993, and there was to be no more subventions as from January 1994.

Group B: This group was made up of six Authorities, namely;

- Anambra-Imo
- Lower Benue
- Lower Niger
- Ogun-Osun
- Upper Benue, and
- Upper Niger.

Thirty-three and one-third per cent ($33^1/_3$%) of their recurrent subvention were to be withdrawn in 1993, the same percentage in 1994, and no more recurrent subvention from 1995.

Group C: This group was made up of three Authorities, namely;

- Benin-Owena
- Cross River, and
- Niger Delta.

Twenty per cent of their recurrent subvention was to be withdrawn each year from 1993-1996, ending up with no subvention in 1997.

By the end of 1995, none of the above groups met their set objectives, however, they all continued their efforts in that direction.

Chapter 9

Floods and Droughts in Nigeria

9.1 Floods and Flood Control

The term "flood" is often explained from a number of perspectives. In hydrology, a flood is any peak flow; this peak flow may or may not overtop the banks of the channel. However, this chapter is concerned with runoff that overflows the channel banks, and inundates its flood plain with the potential to cause damage. In Nigeria, rainfall is the only cause of floods unlike in temperate climates where snow melt can also contribute to the flooding of flood plains.

Flooding at various degrees of severity causing various levels of damage, is an annual event in various parts of the country. However, the historical records of floods are scarce and difficult to obtain. In the majority of cases the serious flood events are reported only in the local newspapers and a few others may be found in scattered official government documents which are usually not easy to obtain.

The first rain gauge station in Nigeria was installed in Lagos in 1892, and this was followed by Calabar in 1900, Sokoto in 1904, and by Ibadan and Kano both in 1905 (Akintola, 1986). The first water-level gauge in Nigeria was erected at Jebba in 1908 in connection with the planned railway bridge across the Niger river. This was followed in 1914 by gauges along the Niger at Baro, Lokoja, Idah, and Onitsha, and along the Benue at Yola, Ibi and Makurdi (NEDECO, 1959).

In 1922, there was an extensive flooding of Sokoto province caused by the floods of Sokoto and Rima rivers. Most of the flood protection works constructed on the banks of Sokoto and Rima rivers upstream from Sokoto were carried away by the flood water and others had to be cut open to allow the flood waters to pass. The Rima floods destroyed a number of villages such as U. Lalle and part of Dan Gerko (KMAG, File No.10708,1923). In 1950, Edjiko stream that later became one of the sources of water for the Wuya-Edozhigi irrigation scheme was heavily flooded when it was overtopped by about 60cm for a greater part of its length (KMAG, File NO. 2702,1947). The flooding of the Kware irrigation project between 1927- 1940 has already been documented in Section 4.3.

The flood of Lake Chad in 1954 caused the lake level to rise to the highest level in 40 to 50 years. The flooding of the lake continued till 1955, and forced the suspension of the construction of the Wulgo dyke for the Wulgo pilot scheme (KMAG, File No. 26902, 1953). A similar severe flooding of the Lake Chad occurred in 1960. Also, the Rima river in 1960 created two-large trenches across the dam of the Wurno irrigation scheme, and flooded the uncompleted scheme. The flood of River Niger of 1974 submerged the rice project in Tada /Shonga in Kwara State for three months thus causing the complete loss of the crop. This particular flood was caused by the high releases of water from Kainji dam upstream of the rice project so as to accommodate in the dam reservoir the expected high flood waters upstream from the dam (Sagua, 1977).

At least two serious dam incidents took place in Nigeria in the 1980s as a result of serious flooding caused by heavy rainfall. The first was the Ogirami dam in Edo State which was commissioned in January 1980 and was overtopped in August of the same year and brought the economic and social life of the people of Akoko-Edo Local Government Area of Edo State to a standstill. The second was the Bagauda dam which failed in August 1988 destroying about 18,000 houses and leaving about 200,000 people homeless; twenty-three lives were claimed to have been lost while two persons were declared missing by the state government. The dam was rehabilitated the same year.

Actions taken to prevent or reduce damage by floods is known as flood control. According to Linsley and Franzini (1964) the commonly accepted measures of reducing floods damage are:

- reduction of peak flow by reservoir,
- confinement of the flow within a predetermined channel by levees, flood walls, or a closed conduit,
- reduction of peak stage by increased velocities resulting from channel improvement,
- diversion of flood waters through by-passes or flood ways,
- temporary or permanent evacuation of the flood plains,
- flood proofing of specific properties, and
- reduction of flood runoff by land management.

Dam building, channel improvement or channelization and the construction of flood protection embankment are probably the most common methods of flood control in Nigeria. The Marabawa Flood Embankment was built in 1926 to protect the Kware irrigation scheme from the Rima river floods. Starting from the first modern dam built in Nigeria in 1923 to 1992, a total of more than 300 small, medium and large dams have been built in the country for various purposes including flood control. These include the Kura (Plateau) dam built in 1929, Ouree (Plateau) 1950, Wurno (Sokoto) 1960, and Kainji (Niger) 1968, Mairuwa (Katsina) 1970, Tiga (Kano) 1975, Pada (Kano) 1980, Jibiya (Katsina) 1990, and Challawa Gorge (Kano) built in 1992, among many others. Dams appear to be the most popular method of flood control in Nigeria.

The government agencies that deal with flood problems in Nigeria are relatively young. The Federal Department of Forestry and the Federal Department of Land Resources were the two most important government agencies involved in soil erosion and flood control until the Federal Ministry of Water Resources was created in 1975 when it took over the studies, design and implementation of flood control projects. The Ecological Fund was created in 1982 to receive the annual allocation of 1 per cent of the Federation Account of Nigeria. In 1983, a National Committee on Ecological Problems was established to review and decide on applications for financing from the Ecological Fund. Out of a total of N1.8 billion committed and released to government agencies between 1985 and 1992, 15 per cent was used for flood control, and 30 per cent for disasters. The other agencies involved in flood control are the Federal Environmental Protection Agency (FEPA) and the Natural Resources Conservation Council.

9.2 Ogunpa Floods

The subject of urban floods is not a concern of this publication. However, because of the notoriety of Ogunpa floods, the subject is given a brief mention in this section. Ogunpa is a small river or stream flowing through the major sections of the city of Ibadan in Oyo State. There have been major floods by this stream in 1948, 1955, 1956, 1960, 1961, 1963, 1969, 1978 and 1980; (*Daily Sketch*, 1980; *Sunday Sketch*, 1980; and Akintola, 1979). The floods of 1956 claimed the lives of 50 people while that of 1960 destroyed more than 500 houses and rendered more than 1000 inhabitants of the city homeless (*Sunday Sketch*, 1980). The flood of 1969 was said to be unique because it resulted from a rainfall of only 25.4mm with a recurrence interval of one year (Akintola, 1979). In 1978 the floods claimed "many" lives and destroyed property "worth millions of Naira"; this flood was caused by a rainfall of 71.4mm.

The Ogunpa flood of 1980 was unprecedented and probably may remain the worst in the history of the city for a long time. As a result of

274.1mm of rain which fell in 12 hours, the whole city of Ibadan nearly collapsed. It was certified that at least 100 persons lost their lives in the flood disaster and 50,000 people were rendered homeless (*Sunday Sketch*, 1980). Several buildings collapsed, the railway tracks were washed away, the Eleyele water works and all the city utilities were severely damaged. The areas of the city seriously affected included Agbogbon, Adamasingba, Bode, Dugbe, Elizabeth Road, Feleye, Gbagi, Molete, Odo - Ona, Oke-Ayo, Omitowoju, Oredein, Osoba, Total Garden and Salvation Army areas.

The Vice - President of the Federal Republic of Nigeria, Dr. Alex Ekwueme, during the presidency of Alhaji Shehu Aliyu Shagari, together with the Oyo State Governor, Chief Bola Ige, undertook an extensive tour of the disaster areas of the city. The Federal Government promised a donation of three million naira to assist the victims of the flood disaster (*Daily Sketch*, 1980; *Sunday Sketch*, 1980).

The development and erection of buildings and other structures in the flood plains and on steep slopes of the city and the indiscriminate dumping of refuse in the Ogunpa stream and its distributaries were held responsible for the regular flooding of the city that almost became an annual event. The channelization of the main valley of the Ogunpa river and other measures to control the regular flooding of Ibadan city was started in 1978. Unfortunately up to the end of 1999 some of these measures had not yet been completed.

9.3 Effects of Flood Control and Other Dams on Downstream Activities.

There is no doubt about the benefits of flood control, irrigation and other dams to the national economy. However, these structures also have their negative impacts. These negative effects particularly on downstream activity are discussed in this section.

The Kainji dam was completed in 1968. As a result of the flood control by this dam, the loss of the fadama land downstream of the dam was said to amount to between 44 and 70 per cent of the pre-dam extent (Adeniyi, 1973). This led to a serious reduction in the farming and fishing activities in the downstream flood plains (Sagua, 1977). There have been similar losses downstream of the Bakolori dam at Birnin Kebbi and elsewhere in the Sokoto - Rima basin. The Bakolori dam was completed in 1978. During the pre-dam period the flood plains were subject to annual flooding. A productive flood recession agriculture based on rice during the wet reason and vegetable crops during the dry season was predominant in the basin. According to Adams (1993), 92 per cent of Birnin Tudu flood plain used to be flooded during the pre-dam period. The dam reduced the flooded area to 33 per cent; the farmers had to change from rice to the less productive soya and millet cropping pattern during the rainy season. The dry season cultivation was completely lost.

Tiga dam which was completed in 1975 reduced the Yobe river flow at Gashua by about 23 per cent; the flooding pattern in the Hadejia-Nguru wetlands was disrupted (Aminu-Kano, 1994). These wetlands in an otherwise arid environment are of great economic and environmental importance. Large quantities of rice, vegetables, fish, livestock and fuelwood are usually produced in the area. In addition to their important role in a groundwater re-charge, the wetlands provide habitats to a great variety of wildlife. Tiga Dam disrupted the functioning of these wetlands and reduced the quantities of products obtained from them. Also, water scarcity downstream from the dam has been exacerbated and sometimes conflicts between the various water user groups were intensified (Aminu - Kano 1994).

Alau dam in the Jere bowl area of Borno State was completed in 1992. During the pre-dam period, the farmers produced rice and vegetables, followed by legume and grain production in the rainy season. Irrigated grains and vegetables and livestock were produced during the dry season. The villages were self-sufficient in food production (Odihi, 1994). However, after the construction of the Alau dam, the frequency, duration, extent and depth of floods in the Jere bowl were reduced and aridity set into the area. The amount of crops produced was said to be less than one-fourth of the pre-dam output. The villages were no longer self-sufficient in food, crime and divorce rates increased in the area and the demographic experience turned from in - migration to out-migration. Proper planning and the participation of all the beneficiary groups in the planning and development process could have averted the Jere bowl experience (Odihi, 1994).

9.4 Droughts

The definition of drought depends on the objective of the definition. Drought could be defined in the context of meteorology, hydrology or agriculture, among others. For example, in meteorology, drought occurs when rainfall is below normally expected values. In agriculture, which is the main concern of this publication, drought occurs when the water supply is insufficient to meet livestock or crop water requirements.

The Sahel and the northern parts of Nigeria have suffered four major droughts during this century. These occurred during the period of 1910 - 1914, 1940 - 1945, 1969-1976 and 1982 - 1985 (Faure and Gac, 1981; Sonuga, 1977; and Nwa, 1993). Faure and Gac (1981) and Sonuga (1977) have indicated that drought maximum seems to occur after approximately every 10 years.

However, they were quick to add that the period of measured data (76 years) is too short to establish a firm cyclicity in the drought appearance. The low rainfall periods exhibited above are not restricted to the sahel and Northern Nigeria alone. The analysis of the rainfall of !badan during the period 1911 - 1971 reflects the pattern of low rainfall periods indicated above (Agboola, 1979).

In addition to rainfall patterns, stream flow and lake levels are also good indicators of the drought conditions. For example, the 1972 - 73 drought reduced the flows of the Kano, Challawa and Watari rivers to between 11 and 56 per cent of the normal flows (Oyebande and Nwa, 1980). In 1908. Lake Chad had a very low level and the whole lake was broken into only two " pools", namely a northern pool and a southern pool. In that year it was noted that villages had sprung up on the dried up lake bed in the northern parts of the lake (Thambyahipillay, 1983). In 1973, as a result of the 1969 - 1976 droughts, the lake was reduced to three small pools while it became a single isolated southern Baga "pool" in 1975 (Thambtahpillay, 1983). The annual minimum daily water-level of Lake Chad at one of its measuring station (Bol), which also reflects the rainfall pattern in the sahel is shown in Figure 9.1.

Apeldoorn (1981) reported incidents of widespread famine in a dozen years since 1900 in Northern Nigeria usually associated with drought. The famine of 1914 in the North was as a result of crop failure in 1913. The famine was particularly severe in Kano and Damagaram The annual rainfall in Sokoto in 1913 was 416 mm and that of Kano was 484 mm, and these low rainfall amounts resulted in crop failure in 1913 that led to the famine of 1914 (KMAG, File No. 10708, 1923). Again, there was famine in the North in 1922 as a result of crop failure in 1921. Although the annual rainfall in Sokoto was 837mm in 1921, heavy rains fell in April and these were followed by an unusually long dry spell; this was what led to the crop failure in 1921 in Sokoto for example, and the famine of 1922 (KMAG, File No. 10708, 1923).

Since the early 1960s, there has been a steady decline in rainfall

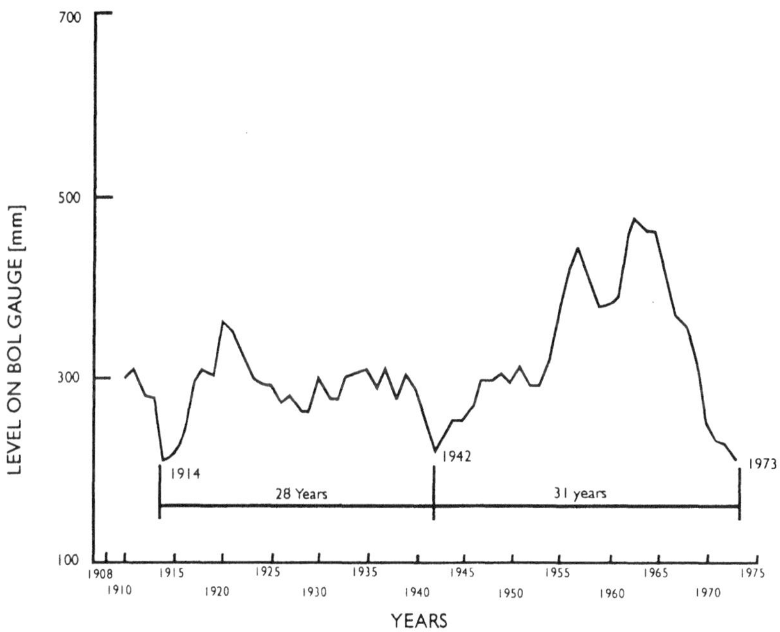

Fig. 9.1: *Annual minimum daily water level of Lake Chad at Bol, 1910 – 1973 (Sonuga, 1977)*

in terms of total amount, amount of rainfall at the peak of the rains and the length of the rainy season throughout the sahel including most parts of Nigeria. This has led to the progressive increase in the occurrence of droughts leading to repeated crop and livestock losses over the last two decades. The declining rainfall trend in Northern Nigeria is illustrated in Figure 9.2 (FAO, 1992; Balogun and Salami, 1994). Drought and the fear of food insecurity have always been strong incentives for governments to pay serious attention to irrigation development in Nigeria.

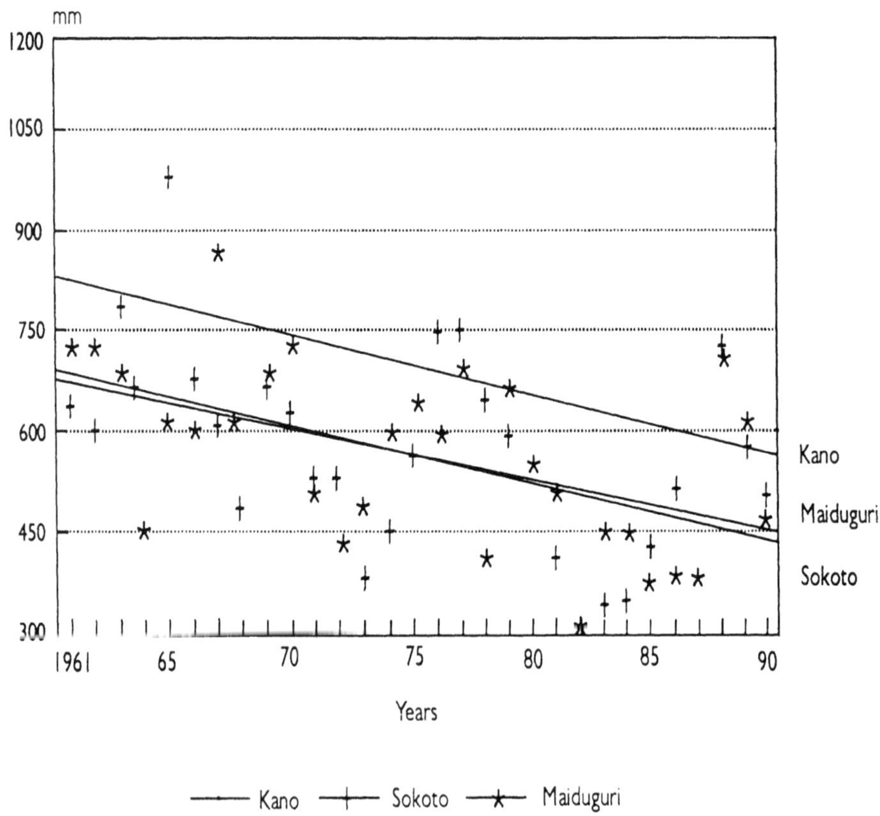

Fig. 9.2: Rainfall trends in Northern Nigeria, 1961-1990 (FAO, 1992)

Chapter 10

Drainage of Agricultural Lands

10.1 Salinity

Salinity is often one of the major problems encountered in irrigated areas and land drainage is the major method of reducing the accumulation of salts in irrigated lands. Due to the serious threat usually posed by salinity to irrigated agriculture, particularly in arid and semi-arid areas, it is often important to identify the existence of the salinity problem early so that the problem could be tackled before crop production is seriously affected.

Because modern irrigation is fairly recent in Nigeria and because the total area under irrigation in the country is relatively small, salinity is not yet a serious problem in Nigeria. However, a number of irrigation schemes have already exhibited symptoms of salinity. The first irrigation in Yau irrigation scheme took place in 1960 and by 1965, 280 ha were actually irrigated; but by 1975 a " considerable" area had been abandoned because of salinity problems. Daya irrigation scheme was first irrigated in 1961. Before 1975, 285 ha of the area was allocated to rice and 675 ha to wheat. By 1975, "most of the wheat area" had been abandoned because of salinity problems. The Abadam scheme was first irrigated in 1964, and 142 ha were under irrigation before 1975 but by 1975 a "considerable area" had been abandoned because of salinity problems (Iqbal, 1975). These three schemes are along the Komadugu - Yobe river in Borno State. Iqbal (1975) also reported salinity problems in parts of Tungun Tudu and Bakura irrigation schemes both in Sokoto State.

In 1962, the Northern Region Ministry of Agriculture and Natural Resources requested the United States Soil Salinity Laboratory, Riverside, California, to survey and report on the salinity state of existing and proposed wheat schemes in Borno, Northern Sardauna, Kano and Sokoto provinces of the North. In addition to Abadam, Daya and Yau schemes mentioned above, the survey found Dankwai lake, Kadandani and Babar Ruga (Tomas) schemes in Kano to be slightly sodium affected (Singh and Maurya, 1979). The 1978 - 79 dry season surveys by the Institute for Agricultural Research (IAR), Zaria found Wurno (Sokoto State), Yamedi (Jigawa State) and Tallum (Adamawa State) schemes with soils indicating salt accumulation on the surface. In the Hadejia basin about 20 - 40 per cent of the older flood plains which are no longer frequently flooded contain soils with alkaline sub-surface horizons. The poorly drained soils (deep, unmottled to slightly mottled, brown sandy loam) of the Kano River Project, contain some salt concentration (Singh and Maurya, 1979).

10.2 Rising Water Table and Water Logging

Other problems that often inflict irrigation schemes around the world include rising water table and water logging. Again, these are problems that are usually controlled through drainage. Just like salinity and for the same reasons given in Section 10.1, rising water table and water logging resulting from irrigation are not yet widespread within irrigation schemes in Nigeria. However, some studies have indicated that these problems are growing in some project areas.

In 1977, Nwa (1982) initiated a study to investigate the pattern of groundwater in the Kadawa sector of the Kano River Irrigation Project (KRIP) Phase I. This phase of the project was expected to equip 22,000 ha of land for irrigation; by 1992, 14,000 ha were actually equipped and irrigated. The first irrigation took place in 1971. Between 1966 and 1967, approximately 500 holes augered within the project area to an average depth of 1.5m were observed during the rainy season (June to September) and only one hole indicated a water level within 1.5m of the ground surface. The study by Nwa (1982) revealed that in 1980 the water table existed in the project area between zero and 60cm below the ground surface at the peak of the rains. It also showed that the water table often rose to within 40cm of the soil surface during the irrigation season. Figure 10.1 shows the pattern of groundwater fluctuation at one of the observation sites during 1980/81. One of the remedies for this problem would be to design and install a sub-surface drainage system for that sector of the project.

Another irrigation project with an evidence of a rising water table is the 5,690 ha Bacita Sugar Estate in Kwara State. The scheme was divided into two sections, one section being irrigated by sprinkler while the other section was by surface irrigation. The first irrigation took place in 1963. Makanjuola and Nwa (1985) showed that the over-irrigation of both the surface and sprinkler irrigated areas was responsible for the noticeable general rise in the water table in both areas.

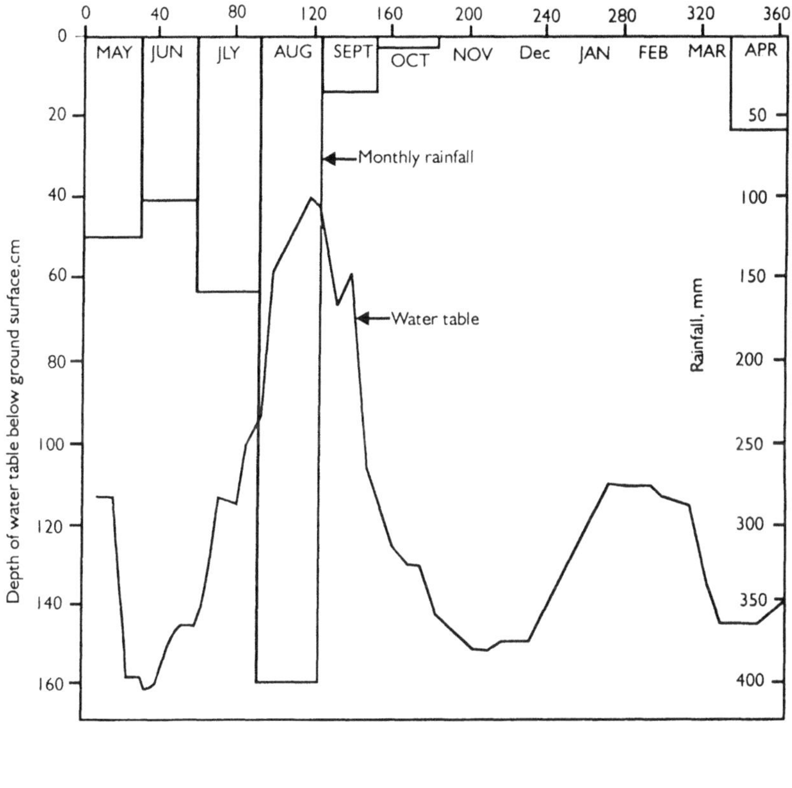

Fig. 10.1: Rainfall and depth of water table below ground. A typical location in Kadawa sector, 1980/81.

10.3 Surface and Sub-Surface Drainage

Surface drainage has always been a part of irrigation schemes. As indicated in Chapter 4, the Kware irrigation scheme of 1925 covered flood control, irrigation, and drainage and land reclamation. The drainage and land reclamation aspect of the scheme involved the channelization of a portion of the Shalla stream and the construction of two branch drains to convey the drainage water from the irrigated area back into the Shalla stream. Apart from the normal drainage of the irrigated land, part of the objective of the Kware drainage works was to lower the water table in the command area and reclaim additional land for irrigation.

In 1952, three of the six village irrigation schemes constructed in Sokoto province were drainage projects. These were the Gummi dam and drainage project, Dangwarema swamp drainage project, and Gawassau swamp drainage project. The Peremabiri Rice Irrigation Scheme of the Niger Delta Basin Development Authority was started in 1967 as a rice polder scheme. Because the scheme was usually subject to water logging from groundwater, river water and rain water, surface drainage was a strong component of the scheme.

As was stated previously, surface drainage has been part of all the irrigation schemes in Nigeria. Unfortunately, sub-surface pipe drainage has not yet been introduced into the Nigerian irrigation system even in schemes which would benefit from such a drainage system. Obviously, the most important limiting factor is the high cost of sub - surface tube drainage system. In an attempt to identify cheap local materials which could be used for sub-surface drainage, Nwa (1981b) experimented with the bamboo (Bambusa vulgaris) in 1979. The result of the study led to the conclusion that the bamboo can be used effectively for field drainage. However, it was recommended that further studies be conducted to establish the drainage design criteria and the drainage life of the material. Since that time not much additional work has been undertaken to promote the bamboo as a field drainage material.

Chapter 11

Irrigation Research, Extension and Training

11.1 Irrigation Research

It is interesting to note that irrigation research was initially a strong component of irrigation development in Nigeria. As stated in Chapter 3, the first irrigation experiment station in Nigeria was the "Baro Experimental Irrigation Farm" started at Baro in November 1908. The objective of the farm was to test the performance of crops such as onion, cotton, tobacco and groundnut during the dry season under irrigation (KMAG, File No. 41577, 1909). In 1914, the Department of Agriculture of the Northern Provinces distributed small samples of wheat and barely for testing and in 1919 the Department issued an improved variety of wheat "Pusa No. 4." This variety became very popular under irrigation in villages such as Gunduse in Kura District of Kano because it out-yielded the local variety (Agriculture Department, Nigeria, 1940).

In 1918, rice growing experiments were conducted at Ahoada in the Delta area of Nigeria by the District Officer and the local schoolmaster. The results were so "impressive" that the researchers saw a good prospect of introducing the crop into other areas of the Delta (Agriculture Department, Nigeria 1918). Similarly in 1921, two imported rice varieties, the Ceylon swamp rice (called "Maliya" by the local people) and Guina ("Creole") were tested in Ilorin against the local rice variety, and both did "much better" than the local variety. The "Maliya" replaced the local variety in the middle part of the Niger valley (Agriculture Department, Nigeria, 1923).

An experimental farm had been established at Maigana (near Zaria) in 1911 but because the farm was small and liable to water logging and the access was poor, it was decided in 1922 that the experimental farm should be transferred to a location nearer to Zaria. It was decided that the Maigana farm should be used as a seed farm. The experimental farm was moved to Shika in 1922. Again, because of poor access, the farm was transferred during the same year from Shika to Samaru. Thus, Samaru became the administrative headquarters of the Department of Agriculture, Northern provinces, and a Regional Research Station in 1922 (Agriculture Department, Nigeria, 1922; Ministry of Agriculture,

Northern Nigeria, 1961). In 1923, Samaru farms were opened and field experiments started in them (Agriculture Department, Nigeria, 1926).

The second irrigation experiment station in Nigeria was the "Kware Irrigation Experimental Farm" which was established within the Kware irrigation scheme in 1926. As indicated in Chapter 4, the farm was used for crop variety selection, and for experiments on time of planting, frequency and amount of irrigation, and land levelling techniques, among others. Between 1926 and 1953 a lot of work was done on selecting the variety of rice and wheat at Kware and several other places. In 1933, Mr. R.R. Glanville was seconded from Sierra Leone to investigate rice growing in mangrove swamps in Nigeria. He located his first experiment which was said to be successful in Warri. Between 1934 and 1942, mangrove swamp rice was grown successfully in Warri, Calabar and Oron but the hard work involved prevented the local people from taking it up "on any scale" (Agriculture Department, Nigeria, 1942).

In 1946, South African varieties of wheat were reported to have done well at Alau near Maiduguri, while 20 new varieties of deep and shallow water rice from South-East Asia were tried out at Bida and Edozhigi in 1951. The variety from Sian was said to be the most attractive (Ministry of Agriculture, Northern Nigeria, 1960; Agriculture Department, Nigeria, 1946). In order to strengthen the work on rice, a Federal Rice Research Station was established at Badeggi in 1953 (Agriculture Department, Nigeria, 1954).

When a self-government was granted to Eastern and Western Regions in 1957, their Departments of Agriculture became Ministries and because of these political changes, the Department of Agriculture of the Northern Region also became Ministry of Agriculture and moved from Samaru to Kaduna. However, Samaru remained as the headquarters of the Research and Specialist Division of the ministry. The division became an autonomous Institute in 1959. When the Ahmadu Bello University (ABU), Zaria, was founded in 1962, this Research Institute was transferred to the University and renamed Institute for Agricultural Research and Specialist Services (IAR&SS) (IAR, 1982). Meanwhile, in 1959, a Hydrological Section of the Irrigation Division of the Northern Region Ministry of Agriculture was set up to collect hydrologic and hydraulic data for irrigation development (FAO, 1966).

When the Federal Ministry of Agriculture and Natural Resources was first created in 1965, its role was formally confined to the administration of agricultural research. Badeggi Federal Rice Research

Station, and IAR&SS in Samaru were the only irrigation related research institute to benefit from the ministry at the time. The two organizations were followed in 1969 by the Institute for Agricultural Research and Training (IAR&T) which was established at Moor Plantation, Ibadan from the former Research and Investigations Division of the Ministry of Agriculture and Natural Resources of Western State of Nigeria. IAR&T was transferred to the University of Ife now Obafemi Awolowo University also in 1969. Although irrigation research was not an important focus of IAR&T, however, the institute conducted research on irrigated crops and studied the water requirements of some of these crops.

In 1970, the Kadawa pilot farm was established to collect research data needed to construct the Kano River Irrigation Project Phase I (KRIP) and in the same year the Nigerian Council for Science and Technology (NCST) was also established to coordinate all scientific research in the country. To sharpen agricultural research, the Agricultural Research Council of Nigeria (ARCN) was established by Decree No. 25 of 1971 to take over the coordination of the research of the agricultural research institutes. This was quickly followed by the Agricultural Research Institutes Decree No. 35 of 1973 which allowed the establishment of research institutes under the aegis of the ARCN.

The Irrigation Research Station, Ngala, which was the first research station of IAR, Samaru, was established in 1972. During the same year, the Board of Governors of IAR directed the institute for the first time to conduct research not only on irrigated crops but on all aspects of irrigation. Following this directive, a part of the existing talents of the institute was diverted to work on irrigated agriculture. The Board of Governors of the institute reviewed the situation of irrigation research in 1974 and directed the institute to strengthen its irrigation research activities. This directive resulted in the establishment of the Soil and Water Management Section of IAR in 1975, the establishment of the Irrigation Research Station, Kadawa also in 1975 and the establishment of Irrigation Research Station, Bakura in 1977 (Iqbal, 1977).

The year, 1975, was a year of special significance to irrigation research in Nigeria. It was early in 1975 that the Federal Department of Water Resources was created and this was followed by the creation of the Federal Ministry of Water Resources later during the same year. Also created in 1975 were the National Cereals Research Institute (NCRI), Moor Plantation Ibadan; the National Horticultural Research Institute

(NIHORT), Ibadan; and the Kainji Lake Research Institute, New Bussa - this institute later became known as the National Institute for Freshwater Fisheries Research (NIFFR). All of the above happened in addition to what took place in IAR, Samaru in 1975. The Lake Chad Research Institute (LCRI), Maiduguri was created a year later in 1976 (Green Revolution National Committee, 1981).

The NCST and ARCN were abolished in 1977 and replaced with the National Science and Technology Development Agency (NSTDA) which became responsible for financing organized scientific and technological research in the country. NSTDA itself was abolished when the Federal Ministry of Science and Technology was established in 1979 to take over its functions, among others (IAR, 1982). The National Water Resources Institute (NWRI), Kaduna was established in 1978. In addition to promoting and developing training courses in water resources, the institute was to serve as a national centre for surface and sub-surface water resources data and to conduct applied investigations into local water resources problems (NWRI, 1984). To undertake research into sugar cane problems, among other functions, the Unilorin Sugar Research Institute (USRI) was established by the University of Ilorin in 1979. Sugar cane is one of the important irrigated crops in Nigeria (Fawole, 1980).

Major irrigation researches are conducted by the national and other research institutes discussed above. However, some aspects of irrigation research are also undertaken by the universities and other establishments in the country. In 1993, there were 24 federal and 11 state universities. Among these were three federal universities of agriculture and 23 faculties of agriculture in other universities. In addition, several of these universities have departments of agricultural, civil, water resources and environmental engineering; geography; geology; applied geophysics; meteorology; agronomy and water management in addition to the departments of management sciences. Several of these departments and units have undertaken one form of irrigation research or the other. These departments and units are summarized in Table 11.1 after Nwa (1993).

Table 11.1 Number of departments of various disciplines in Nigerian Universities (March 1993)

S/No.	Disciplines	Number of Departments
1.	Agricultural Engineering	14
2.	Civil Engineering	22
3.	Water Resources and Environmental Engineering	1
4.	Geography	25
5.	Geology	15
6.	Applied Geophysics	3
7.	Meteorology	3
8.	Agricultural Meteorology and Water Management	1
9.	Agronomy	3
10.	Soil Science	Several but as part of agriculture

According to Nwa (1983, 1993), the best organized research programmes in support of irrigation are in the area of irrigated crops improvement. The most important irrigated crops include rice, wheat, maize, sugar cane, cowpea and vegetables, among others. Some of the research institutes have the mandate to improve these and other irrigated crops. The other aspects of irrigation research are less organized than irrigated crops research. Fewer research institutes have the mandate to undertake other aspects of irrigation research (Nwa, 1983; 1993).

11.2 Irrigation Extension

Irrigation extension, in one form or the other, has always been a component of irrigation development in Nigeria. The principal method used over the years has been "demonstration." To introduce a new irrigated crop into an old or new irrigation area or to expand an irrigated crop into new areas, research was usually conducted to select the best crop varieties and then demonstration plots were established to teach the farmers how to grow these crops. This was the case when rice, wheat and other irrigated crops were introduced or their production expanded in various areas. The "Baro Experimental Farm" of 1908 and the Kware Irrigation Experimental Farm" of 1926 were used for both research and demonstration purposes.

In 1927, the National Department of Agriculture was reorganized. In the reorganization, it was decided to appoint two Assistant Directors of Agriculture for the department in addition to the Director. One Assistant Director was responsible for the Southern provinces while the other took charge of the Northern provinces. Each Assistant Director was provided with a Deputy. The Assistant Directors were responsible for all the experimental and extension work in their groups of provinces while the Director took charge of other administrative responsibilities at the headquarters. This arrangement was planned in 1926 and it actually took effect in 1927 (Agriculture Department, Nigeria, 1926).

As a matter of policy, the Department of Agriculture decided to establish "Farm Centres" to serve as demonstration, general headquarters, and centres of extension work for each administrative district. The first Farm Centre was established in 1931 at Kafin Soli near Katsina while the second was established at Maigana near Zaria during the same year. Thus, 2 of these centres were established in 1931, 7 in 1932, 5 in 1933 and 3 in 1934 in Kano, Sokoto and Zaria provinces (Agriculture Department, Nigeria, 1933). Although the majority of these centres provided extension work to rain-fed agriculture, a few of them did serve irrigated agriculture.

By 1934, it had already been demonstrated at Kware by the Department of Agriculture that the use of bullocks to puddle the irrigated rice fields was much more profitable than broadcasting the rice. The problem was how to convince the farmers to adopt the technique in view of the hard work involved. One technique tried by the Department of Agriculture in 1934 at Kware was to select three men, one of them an ex-middle school boy, and provide each of them with land, a pair of bullocks, labour and other inputs. Although these men were paid as labourers, they were allowed to manage the farms on their own. The Department kept a record of the cost of production and returns. Their operations further confirmed that the system was profitable but it did not solve the problem of adoption (Agriculture Department, Nigeria, 1934).

Although Samaru became the headquarters of the Northern

Department of Agriculture and Regional Research Station in 1922, it was only in 1963 that the Ministry of Agriculture of the North established the Extension and Research Liaison Section (ERLS) as part of its Field Services Division stationed in Samaru. When Nigeria was broken into 12 states in 1967, each of the six states in the North created a Ministry of Natural Resources and Cooperatives and ERLS was merged with IAR&SS in Samaru. It was in 1975 that ERLS was separated from IAR&SS and became known as the Agricultural Extension Research Liaison Services (AERLS) (Green Revolution National Committee, 1981). AERLS became national in 1987 and modified its name to National Agricultural Extension Research Liaison Service (NAERLS). This agency became the leading irrigation extension agency in Nigeria since its creation. One of its contributions was the organization of the 1st Nigerian Irrigation Seminar in 1971 "to bring together people from all the Northern States representing the Ministries, Consultant Companies, FAO, and the ABU complex that are interested in irrigation" (IAR, 1991). ERLS ran the annual seminar for three years before it became national. In 1974, the seminar was jointly organized by AERLS and the Federal Department of Agriculture (Land and Water Resources Division). When the Federal Department of Water Resources and the Federal Ministry of Water Resources were created in 1975, the Ministry became involved in the organization of the seminar with AERLS through the Federal Department of Water Resources.

When the River Basin Development Authorities (RBDA) were created between 1973 and 1976, each RBDA took the responsibility of providing irrigation extension services to its farmers. This situation continued until 1987 when the authorities were divested of their direct agricultural production functions. The extension services to the river basin farmers were supposed to be provided by the Agricultural Development Projects (ADPs). The ADPs were already responsible for providing extension services to the rain fed farmers. However, because the ADPs had very little experience in irrigation extension, many of the RBDAs argued against handing over irrigation extension to the ADPs. Up to the end of 1995 the official policy of handing over irrigation extension to the ADPs had not been changed, but the debate also continued (Nwa, 1994c).

11.3 Irrigation Training

During the early years of irrigation development, the training of the staff for irrigation work was completely informal. For example, in 1926, some staff of the Agriculture Department (North) were trained at the Kware Irrigation Farm on how to use the bullock for puddling and levelling rice fields. When the irrigation engineer, John Kitching, was first appointed to Sokoto in 1949, as his first task, he arranged a training course for the junior local staff. With the assistance of the Survey Department, he organized a training school which provided a sequence of six-month courses on surveying and other irrigation works, starting with the first set of 12 trainees (Agriculture Department, Nigeria, 1950). The school continued until it was closed down in 1954 since enough staff at that time had been trained for Sokoto (Agriculture Department, Northern Region, 1954). However, the problem of shortage of trained staff in all aspects of agriculture in Northern Region was still serious, and therefore to tackle this problem, the Northern Region Ministry of Agriculture created a new division, the Agricultural Training and Education Division in 1957. The main functions of this division were to train lower grade, intermediate and upper intermediate technical staff at local training schools for the ministry; to arrange overseas in-service training courses for the staff; and to train young men to become progressive farmers (Ministry of Agriculture, Northern Region, 1965). To address the technical staff problems in the irrigation sector, the Northern Ministry of Agriculture established an "Irrigation School" in Sokoto in 1964 with a first batch of eight students. The school ran a two-year course training the students for the posts of Technical Assistants Grades I and II (Ministry of Agriculture, Northern Region, 1965).

To fill the gap between the trainees from the School of Irrigation, Sokoto and trained people who could effectively advise farmers and others on irrigation water management it was decided to start an Irrigation Agronomy Course at the Federal Soil and Water Conservation Training Centre, Jos. So as to take care of the needs of the whole Federation, the course was started by the Federal Ministry of Agriculture and Natural Resources in 1970 with 17 students. To qualify for admission into this course, it was agreed that second-year students from Schools of Agriculture such as those at Samaru, Kabba, Umudike, etc. be transferred to Jos for the course (ERLS, 1971).

Meanwhile, the plans to construct large-scale irrigation schemes in Kano and Sokoto brought pressure on these states to establish training programmes for the junior staff that would be needed in these schemes. Consequently in 1972 the Kano State Ministry of Agriculture and Natural Resources started a 2-year "Irrigation Instructors' Course" at Kadawa, near Kano. For entry into the course the candidates needed Primary VII or Secondary I, II or III qualification. This was followed in 1978 by a "School of Irrigation and Extension Services" started by the Sokoto-Rima Basin Development Authority at Bakura near Sokoto. Initially, primary school and Secondary III leavers and grade II teachers were admitted into the school but the entry requirements were later raised to secondary school leavers and those who had attempted the West African School Certificate examination. During this period also, the "Irrigation School" in Sokoto "was almost running out of students." Because of this, Kaduna Polytechnic took over the courses of that school in 1972 (ERLS, 1972; AERLS, 1981).

At the same time, a number of polytechnics and similar institutions were being established all over the country and several of these institutions included irrigation training in their various programmes. The polytechnics and similar institutions which included irrigation training among their programmes are shown in Table 11.2. The disciplines in which first and postgraduate degrees in some aspects of irrigation are obtained in Nigerian universities are given in Table 11.1.

Table 11.2: Polytechnics and similar institutions that offer irrigation courses (1985)

S/No.	Institution	Course	Diploma	Year Founded
1.	Federal Polytechnic, Kaduna	Agric. Mechanization/ Engineering	HND	1968
2.	Kwara Polytechnic, Ilorin	Agric. Engineering and Water Resources Engineering	ND, HND	1968
3.	Ramat Polytechnic, Maiduguri	Irrigation	ND, HND	1976
4.	The Polytechnic of Sokoto State, Birnin Kebbi	Irrigation	ND, HND	1976
5.	Lagos State College of Science and Technology, Ikeja	Agricultural Engineering	ND, HND	1977
6.	National Water Institute, Kaduna	Water Resources Engineering	ND, HND	1978
7.	College of Technology, Owerri	Agricultural Engineering	ND/HND	1978
8.	Federal Polytechnic, Bauchi	Agricultural Engineering	ND, HND	1979

Note: ND = National Diploma; HND = Higher National Diploma
Source: NBTE, 1985.

Chapter 12

Land Tenure, Water Laws, Rights and Customs

12.1 Land Tenure

Awogbade (1982) defined land tenure as the manner in which land is acquired, owned, used and transferred. This definition contains the three principal components of land tenure defined by Adegboye (1976) which are ownership, transfer and use. Before the colonial period, land tenure in the area now called Nigeria was determined by customary laws which varied somewhat from ethnic group to ethnic group. But basically land was owned by the individual, family and the group. The Clan Head, Chief, Oba or Emir held land in trust for the family or community (Awoyemi, 1982). The customary land tenure in the various areas of the country has not always been unchanging. In the North for example, since the land belonged to all the people, the individual rights to the group land could not initially be subject to any commercial transactions such as rent, mortgaging or sale. But as land became less available, even peasants could sell, pledge or lease their lands and their children could inherit the lands (Raza and Abdullahi, 1979; Saidu, 1994). This is why fragmentation of lands has become a serious problem for the development and management of large-scale irrigation schemes today.

Because land was communally owned in Nigeria before 1914, no individual possessed "Fee simple" or absolute title to land. After the amalgamation of the Northern and Southern Protectorates in 1914, "Fee simple ownership", supplemented by a 99-year lease of land was established in Lagos. This was extended to other parts of the country later (Aladejana, 1982). But before this, the Lands and Native Rights proclamation had been issued for the North in 1910 putting all the lands of Northern Nigeria in the hands of the state. The Native Land Rights Ordinance of 1916 amended the proclamation of 1910 and was based on the theory that the Emirs owned all the lands in the North. After independence in 1960, the Ordinance of 1916 was made into law in 1962. This law put all the lands in Northern Nigeria under the control of the government to be "held and administered for use and common benefit of the native" and "no title to the occupation and use of any such lands shall be valid without the consent of the governor." The governor, in exercising his authority was advised to have some regard for the laws and customs of the local people. The government was, therefore, given the legislative control while the traditional rulers were given the customary administrative control over land matters. The peasant cultivators were given unreserved usufructuary rights. According to Raza and Abdullahi (1979), these laws had little or no effect on the people of the North, as they continued to inherit, sell, pledge or lease land as before.

To simplify the processes of obtaining land for agricultural and other purposes, among other important reasons, the Land Use Act 1978, enacted as the Land Use Decree No.6 of 1978, came into force on 29th March, 1978 and was made part of the Nigerian constitution (Awoyemi, 1982; Lot, 1982; Famoriyo, 1982; Nigeria, 1978). The first part of the Act states as follows:

1. Subject to the provision of this Decree, all land comprised in the territory of each State in the Federation are hereby vested in the Military Governor of that State and such shall be held in trust and administered for the use and common benefit of all Nigerians in accordance with the provisions of this Decree.
2. As from the commencement of the Decree,
   a) all land in urban areas shall be under the control, management of the Military Governor of each State; and
   b) all other land shall, subject to this Decree, be under the control and management of the Local Government within the area of jurisdiction of which the land is situated.

In effect, this law was meant to bring all of Nigeria under the same land law since it was similar to the law that was made for Northern Nigeria in 1962. Like in the North before it, the Act did not seem to change much in the customary land practices in the South (Awoyemi, 1982).

12.2 Water Laws, Rights and Customs

In many parts of the world anyone whose land is alongside a river or stream owns half the riverbed and is entitled to use half the water on his land. This is known as "riparian rights" (Schultz International Limited, 1975). Since "riparian rights" were the common law in England and since the common law of England was to prevail in Nigeria during the colonial period, "riparian rights" should have been introduced into

Nigeria. But because the concept of right to land including a right to water is strange to Nigeria and this concept is fundamental to "riparian rights" this is why "riparian rights" have not been applied in Nigeria. In Nigeria, there has been no law of water until recently; each person was free to use whatever water was available for his purposes.

However, attempts have always been made to charge farmers for the water they use in irrigation schemes. In Kware in 1927, the farmers did not pay for the water initially but there were discussions already that the farmers would have to pay some water rate in the future. In 1956, the irrigation farmers at Kware were paying 5, 8 and 10 shillings, depending on the "amenities" offered to the various parts of the project (KMAG, File No. IRR 104, 1958). The managers of several of the early irrigation schemes tried to declare the schemes "Native Authority Settlement Area" to give them the authority to collect water and other charges and to give them the power to enforce the rules governing these schemes. In spite of early attempts to do so, it was not until 1963 that Kware was declared a settlement area.

One of the irrigation regulations to be enacted was the "Bida Native Authority (Irrigation) Rules, 1956" which was promulgated by the Bida Native Authority and approved by the Executive Council of the Northern Region in 1956 for Edozhigi, Badeggi and Toroko irrigation schemes. The most important purpose of the regulation was to fix the annual irrigation rate for each scheme and collect these charges. The regulation also gave the Native Authority the power to suspend water supply and specified the penalties against those who failed to comply with the rules (KMAG, File No. 49845, 1956).

Recently "Irrigation Bye-Laws 1990" was approved for all the River Basin Development Authorities by the National Council on Water Resources (NCWR, 1990). This By-Law also deals mainly with irrigation and other water charges. The irrigation water rate of N500 per hectare per season was approved for all crops. This rate is subject to review by the Authority without notice. The by-law also provided for a penalty of ₦1,000/ha for unauthorized use of water and ₦2,000 for damage to irrigation works.

One of the important achievements of the Federal Ministry of Water Resources and the National Council on Water Resources was that for the first time a Water Resources Law was passed for Nigeria. This law took about 20 years to prepare and was promulgated in 1993 as "Water Resources Decree 1993" (Nigeria, 1993). Among several other things the Decree:

1. Vests the rights to the use and control of all surface and groundwater in the Federal Government.
2. Approves "riparian rights" for all Nigerians.
3. Approves the rights to acquire and use water for all Nigerian or any public authority in Nigeria.
4. Gives power to the Secretary (Federal Minister of Water Resources) to, among several other things:
   a) regulate and control the use of surface and groundwater
   b) ensure adequate supply of suitable water for all purposes including irrigation, and
   c) draw up from time to time an up-to-date comprehensive masterplan for the development, use, control, protection, management and administration of all water resources.

Chapter 13

International Organizations, Treaties and Agreements

13.1 International Commission on Irrigation and Drainage (ICID)

The International Commission on Irrigation and Drainage (ICID) is the leading Non-Governmental Organization (NGO) in the field of irrigation, drainage and flood control in the world. ICID was established in India in 1950 with 10 countries which were Egypt, India, Indonesia, Iraq, Israel, Italy, Netherlands, Sri Lanka, Thailand and Yugoslavia; its headquarters is in New Delhi, India.

The objectives of ICID are to stimulate and promote the development and application of the arts, sciences, and techniques of engineering, agriculture, economics, ecology and social sciences in managing water and land resources for irrigation, drainage, flood control and river training and/or for research in a more comprehensive manner adopting up-to-date techniques (ICID, 1995). The Commission accomplishes its objectives by:

- conducting studies — collection and analysis of data/ information;
- technology transfer — dissemination of technology and experiences through issuance of technical publications;
- organizing global technical congresses, regional conferences, workshops, and round tables; and
- cooperating with other international organizations.

The first five Presidents of ICID were Dr. A.N. Khosla (India; 1950-1954); Mr. H. Suleiman (Egypt; 1954-1957); Mr. W.A. Dexheimer (U.S.A; 1957-1960); Mr. N.D. Gulhati (India; 1960-1963); and Professor P. Danel (France; 1963-1966). So far the Secretaries - General have all come from India and these were Mr. N.D. Gulhati (1950-1957); Mr. Yadava Mohan (1957-1963); Mr. K. K. Framji (1963-1988); Dr. R.S. Varshney (1988-1992); and Dr. M.A. Chitale who assumed office in 1993.

In 1956, six years after the establishment of ICID, the Department of Agriculture, Northern Region (Nigeria) applied and was admitted into the Commission. When the Department became the Northern Region Ministry of Agriculture in 1957, the Ministry continued to represent Northern Nigeria in ICID. In 1967, Nigeria was broken into 12 states, the six states of the North being North-Western State with the headquarters in Sokoto, North-Central State (Kaduna), North-Eastern State (Maiduguri), Kwara State (Ilorin), Benue-Plateau State (Jos) and Kano State (Kano). The Ministry of Agriculture of the North became the Ministry of Natural Resources (MNR) in each of the six states. Because of these changes the North withdrew its membership of ICID in April 1968 and in July of the same year, the Ministry of Natural Resources, North-Western State (Sokoto) wrote to the Permanent Secretary, Federal Ministry of Agriculture and Natural Resources (FMANR), Lagos and requested the Federal Ministry to take over the responsibility of ICID on behalf of all the State Ministries of Agriculture. Following the letter from Sokoto, all the six MNR from the North met and endorsed the position of MNR, Sokoto. Consequently, the Federal Ministry of Agriculture and Natural Resources agreed to take over the responsibility of ICID on behalf of all the Ministries of Agriculture in Nigeria. FMANR was advised by the ICID headquarters in New Delhi to make a fresh application for Nigeria to become a member of ICID. After all the formalities, Nigeria was admitted as a member of ICID in 1970 and FMANR acted as the Nigerian National Committee of ICID. When the Federal Ministry of Water Resources (FMWR) was created in 1975 it took over the responsibility for ICID (FMANR, File No. AGR 137/S.1, 1968). Since 1970 the Nigerian National Committee has been a very active member of ICID. In 1995 there were 66 active and 17 inactive member-countries of ICID (ICID, 1995).

In 1960, the International Executive Council of ICID during its meeting in Madrid, Spain, approved the principle of regional technical conferences to discuss technical subjects of particular interest to the countries of the region. Consequently, the first Afro-Asian Regional Conference of ICID was held in Tashkent, USSR in 1976. The second conference followed in 1978 in Manila, Philippines, while the third conference was held in New Delhi, India in 1980. Following an invitation by the Nigerian National Committee to ICID, the fourth Afro-Asian Regional Conference of ICID was held in Lagos in 1982 on the theme "River Basin Development for Food Production." A total of 201 delegates from 20 countries of Africa, Asia, Europe and North America participated in the conference. The countries represented were Algeria, Bangladesh, Benin, Czechoslovakia, Greece, India, Iraq, Italy, Cote

D'lvoire, Japan, Kenya, Malaysia, Nigeria, Pakistan, Philippines, Thailand, United Kingdom, U.S.A., U.S.S.R. and Zimbabwe. Also represented were five international organizations, namely, FAO, WHO, UNESCO, Niger Basin Authority and International Institute of Tropical Agriculture (IITA). The officers of ICID at that time who attended the conference were:

Mr. W.R. Rangely (United Kingdom)	-	President, ICID
Mr. G. Papadopoulos (Greece)	-	President, Honoraire, ICID
Professor M. Holy (Czechoslovakia)	-	President, Honoraire, ICID
Mr. K.K. Framji (India)	-	Secretary-General, ICID
Mr. N.O. Popoola (Nigeria)	-	Vice-President, ICID

Because of the active participation of Nigeria in ICID, Engr. N.O. Popoola was elected a Vice-President of ICID from 1980-1983 and Professor E.U. Nwa was also elected a Vice-President from 1984-1987. Unfortunately, the declining national economy adversely affected the ability of Nigeria to actively participate in ICID affairs from 1990. To reverse this trend an attempt was made to organize the National Committee as a Non-Governmental Organization. To this end, a meeting of members was called by the Director of the Federal Department of Irrigation and Drainage, Engr. Inuwa K. Musa, at the Lower Niger River Basin Development Authority, Ilorin in 1994. At that meeting the synonym, NINCID (Nigerian National Committee on Irrigation and Drainage), was adopted for the National Committee, new officers were elected and a new constitution was adopted. Engr. I.K. Musa was elected as Chairman; the three Vice-Chairmen elected were Engr. A.A. Lawal, Professor E.U. Nwa and Engr. N.O. Popoola. The other members of the Executive Committee were Engr. M.S. Adamu, Professor F.A. Adeniji, Dr. A.A. Ramalan, Dr. S.M. Toro and Mr. S. Abetianbe.

13.2 Niger Basin Authority (NBA)

River Niger is one of the three or four most important rivers in Africa flowing through several countries. The desire to develop the River Niger basin encouraged the riparian countries to seek international cooperation among themselves. Therefore, in 1963, nine riparian countries met in Niamey, the capital of Niger Republic and agreed to create the River Niger Commission (RNC). This agreement was signed as an Act of Niamey in 1963. The nine countries which signed the Act were:

- Benin,
- Burkina Faso,
- Cameroon,
- Chad,
- Cote D'Ivoire,
- Guinea,
- Mali,
- Niger, and
- Nigeria.

The aim of the Commission was to promote cooperation between member-states and to develop the resources of the basin particularly in the fields of energy, fisheries, forestry, transport, communication and industry (FMWR, File No. FDWR/WR/S.57, 1985). The headquarters of the Commission is at Niamey. Mr. E.A. Dehinde of Nigeria was the first Executive Secretary of the Commission from 1964-1979. In 1975, the Republic of Chad withdrew from the Commission because River Niger does not flow through her territory. During the meeting of the Heads of States of the member-countries of the Commission in 1980 in Conakry and Faranah both in Guinea, they decided to change the name of the Commission to Niger Basin Authority (NBA) and also appointed Mr. Diawa Mory Traore as the second Executive Secretary of the Authority.

13.3 Lake Chad Basin Commission (LCBC)

The Lake Chad Basin is made up of parts of the four countries of Cameroon, Chad, Niger and Nigeria. On 22nd May, 1964, these four countries signed a convention creating the Lake Chad Basin Commission (LCBC) to ensure a rational and equitable development of the water and other natural resources of the basin to the benefit of the member-states of the Commission. This convention was signed at Fort Lamy (now Ndjamena), the capital of Chad Republic which serves as the headquarters of the Commission (LCBC, 1973). The Central African Republic which is the source of Shari, the most important river feeding the Lake Chad, became a member of the LCBC in 1986.

The area of the hydrographical basin of Lake Chad is said to be 3,500,000 km^2 while that of the conventional basin is 427,300 km^2 (LCBC, 1990). The "hydrographical basin" refers to the natural catchment (watershed) of the lake while the "conventional basin" refers to the area arbitrarily defined by the member-countries of the LCBC for development purposes. The distribution of the conventional basin among member-states of the Commission is shown in Table 13.1 and the development of irrigation within the basin by member states of the Commission is given in Table 13.2.

Table 13.1: Distribution of the conventional basin of Lake Chad among member-countries of LCBC

Country	Areal extent of conventional basin km^2
Chad	178,300
Niger	121,500
Nigeria	90,000
Cameroon	37,500
Total	427,300

Source: LCBC, 1990.

Table 13.2: The situation of irrigation in Lake Chad Basin in 1980 (Unit: ha)

Country	Location of Scheme	Name of Scheme	Operational	Under Construction	Studied Scheme	Total Construction	Maximum Development
	A1 North East Polders	Bol	12,500			12500	60,000
	A2 Banks of Chari	P.I.C	100		1000	1,100	10,000
	A3 R.B. of Logone	Safegui-Deressia	100	5,000	5,000	11,000	25,000
CHAD	A4 R.B. of Logone	Casiers de Bongor (SEMRY 1)	2,500			2,500	15,000
	A5 Lower Logone-Lower Chari Malo		100		5,300	5,400	12,500
– –			15,300	5,000	12,200	32,500	142,500
	B1 L.B. of Logone	SEMRY 1	5,300			5,300	10,000
	B2 L.B. of Logone	SEMRY 2		6,300		6,300	10,000
B CAME-ROON	B3 Lower Logone-Lower Chari	Kousseri	200		5,000	5,700	12,500
	B4 Mayo Tanga				7,200	7,200	7,200
			5,500	6,300	12,700	24,500	46,700
	C1 Yadeeram				35,000	35,000	35,000
	C2 Nagadda	Jere Bowl			9,000	9,000	9,000
C	C3 Ebeji		1,200			1,200	2.000
NIGERIA	C4 S.C.I.P.			47.000	17,000	64,000	64,000
	C5 South-West Polders	Baga-Kirenowa			30,700	30,00	90,000
	C6 R.B. of Komadougou	(Diffa)	1,700			1,700	4,000
– –			2,900	47,000	91,700	141,600	204,000
D NIGER	DI L.B. of Komadougou	(Diffa)	100		2,300	2,400	2,400
All Countries		23,800	58,300	118,900	201,000	368,600	

Source: Nwa, 1993a

13.4 Nigeria-Niger Joint Commission for Cooperation (NNJC)

Nigeria and Niger have four major drainage basins common to both countries. These are:

- C Gada-Goulbi of Maradi,
- C Komadugu-Yobe,
- C Maggia-Lamido, and
- C Tagwai-Elfadama.

In the early 1970s, the Nigeria-Niger Joint Commission (NNJC) was established in Niamey, the capital of Niger Republic where the headquarters of the Commission is based. Under the bilateral agreement, NNJC was vested with the responsibility of ensuring an effective and mutually beneficial cooperation between the two countries in scientific, technical and economic matters. In particular, the Commission was to monitor and recommend development options within the four common basins especially in the area of water extraction and development.

In 1982, NNJC was able to secure financial assistance from UNDP to conduct studies on the hydrology, hydrogeology and socio-economy of the four common basins; the studies were aimed at characterizing the development potentials and constraints of the basins.

Chapter 14

Current Problems of Irrigation, Drainage and Flood Control in Nigeria

14.1 Problems of the Irrigation and Drainage Sub-Sector

The immediate past and current problems of the irrigation and drainage sub-sector are many and varied. Some of these problems are summarized below.

1. *Deficiencies in Project Conceptualization and Design*

    - the review of some previous project feasibility reports indicated that the majority lacked proper preparation of topographic maps, and soil and land classification surveys; in some cases the hydrological information were misinterpreted.
    - there was the tendency to adopt levels of technology that were beyond the absorptive capacity of the ultimate users, thus rendering the projects unsustainable.
    - little or no efforts were made to design projects for multipurpose use for irrigation, water supply and hydropower, among other uses; this concept could reduce the cost of the projects for each sector and make them more sustainable.

2. *Inconsistency of Government Policy*

There is not yet a comprehensive irrigation policy in Nigeria, a comprehensive irrigation policy is now at its second draft stage but worse still is the frequent changes in existing policy. For example, it is not yet quite clear whether irrigation extension should be handled by the irrigation agencies or by the States Agricultural Development Projects (ADPs) who are the official organs for the agricultural extension services. It has been argued that since the ADPs do not have the experience and the manpower to undertake irrigation extension, the service should be handled by the irrigation agencies. In any case, the future of the ADPs has not yet been assured.

3. *Inadequate Funding and Untimely Release of Approved Finances*

Perhaps the greatest problem confronting irrigation development in Nigeria today is poor funding. In addition to the inadequate funding, the release of the approved finances has often been untimely and incomplete. This has contributed to the high cost of projects in a country where the rate of inflation is very high.

4. *Poor Operation and Maintenance*

Poor operation and maintenance are largely responsible for the poor performance of several of the irrigation schemes in Nigeria. The major factor responsible for the poor operation and maintenance is inadequate funding of the projects. However, other important factors include the low priority often attached to the project maintenance especially new projects, and the inability of some project managers to involve the farmers in the operation and maintenance of their schemes.

4. *Poor Performance of Established Projects*

Some of the established projects show poor performance in the form of:
- decreasing crop yields
- low return on investment
- increasing salinity
- rising water table, and
- water logging, among others.

The majority of the above difficulties arise as a result of poor operation and maintenance which themselves, are largely due to inadequate finances.

6. *Problems of Farm Input and Farm Support Services*

Tractor service, improved seeds, fertilizers, chemicals, and similar farm inputs have not always been readily available to the projects and the farmers. In addition, problems of such support services as credit and marketing have helped to lower the performance of some irrigation projects in the country.

7. *Land Tenure Problems*

In some parts of Nigeria land belongs to the families while in others land is held in trust by the government for the communities. In spite of the Land Use Act of 1978, the practical effect of the various land tenure systems is that in some areas, it is not easy to acquire land for irrigation projects. Even where land is expected to be developed and handed back to their owners for farming, the existing land fragmentation has often resulted in inefficient irrigation and farm machinery operations.

8. *Weak Extension and Research Support*

Extension and research support are essential components of efficient irrigation development. The weaknesses of these services have contributed to the slow rate of development and the poor performance of irrigation in Nigeria.

9. *Slow Rate of Irrigation Development*

The rate of developing formal irrigation in Nigeria has been very slow indeed. Table 14.1 shows that Oyan and Zobe dams were completed in 1983 and yet the infrastructural development for downstream irrigation has hardly been started. The construction of Tiga dam started in 1971 for the purpose of developing 22,000 ha of the Kano River Project Phase I and, in 1992, 21 years later, 14,000 ha of the project were completed. Unfortunately, some of the projects need to be rehabilitated before their construction is completed.

10. *Developed Irrigation Headworks Without Downstream Development*

A large number of irrigation dams have been completed but without or with minimal infrastructure for downstream irrigation. Some of these dams may have lost a large percentage of their storage capacity due to siltation. A list of some of these dams is given in Table 14.1. In 1995, $3,760 \times 10^{6}$ m^3 of water was said to be available in existing reservoirs for irrigation and only 350×10^{6} m^3 was actually used for irrigation (NWRMP, 1995).

11. *Absence of Farmer Participation*

The incorporation of farmers into the management, operation and maintenance of irrigation schemes is now at an experimental stage in Nigeria. Water Users' Association (WUAs), farmer cooperatives, and similar farmers' organizations have contributed greatly to the operation and maintenance of irrigation projects in other places.

12. *Lack of Skilled and Experienced Manpower*

Inadequate skilled and experienced technical manpower has been recognized as one of the major constraints to sustainable irrigation development in Nigeria. Apart from limiting the capacity to evaluate consultants' and other technical reports, this problem has also limited local initiatives at operating, maintaining and managing the projects.

Table 14.1 Some completed irrigation dams with minimal or without downstream irrigation infrastructure (1992)

S/No	Name of dam	RBRDA	Year completed	Proposed Irrigation area ha	Area equipped for irrigation ha	Gross storage m^{m3}
1.	Oyan dam	Ogun State	1983	12,000	-	270
2.	Zobe dam	Sokoto Rima	1983	8,137	100	177
3.	Goronyo	Sokoto-Rima	1984	17,000	-	942
4.	Ero dam	Benin-Owena	1987	2,000	60	20
5.	Dadin kowa	Upper-Benue	1988	44,000	150	2,855
6.	Doma dam	Lower-Benue	1988	20,000	-	38
7.	Jibiya	Sokoto-Rima	1990	3,450	-	142
8.	Ikere Gorge dam	Ogun-Oshun	1991	12,000	-	690
9.	Alau dam	Chad Basin	1992	8,000	-	112

13. *Lack of Sufficient Awareness of the Need for Irrigation in the Southern States*

Although the potential evapotranspiration is usually greater than rainfall during four to five months of the year even in the very wet areas of the South, the awareness of the need and advantages of irrigation in the southern states has been fairly recent at the federal, state and local government levels even among engineering, agricultural and such professionals. This is why local initiatives were insignificant until the RBDAs were extended to the South more for political than technical reasons (Nwa, 1992).

14. *Lack of Irrigation Experience by Farmers in the South*

Irrigation is a new technology to the farmers in the South who have been used to rain-fed farming. It is not an easy task to get the farmers to change their way of life which is entailed if dry season farming (irrigation) is to be combined with rain-fed agriculture.

15. *Lack of Reliable Data Bases*

As long as adequate and reliable water resources data are not available, planning, design and management of irrigation projects will continue to remain a guesswork. Adequate, accurate, climatic, hydrologic, hydrogeologic, irrigation and other associated natural resources data are not available in the country. Reliable data bases of appropriate length of record are essential for sustainable irrigation development and management.

16. *Lack of Environmental Impact Assessment of Projects*

Health problems have already developed in some irrigation projects in Nigeria. Some watersheds in which dams have been constructed are said to be deteriorating at an alarming rate. There appears to be no evidence of any systematic environmental impact consideration carried out in any of the irrigation projects (NWRMP, 1995).

The environmental aspects of irrigation development involves:

- public health
- fisheries
- aquatic weeds
- consideration for wildlife
- downstream hydrology
- agricultural pest management
- increased use of fertilizer
- impacts on groundwater
- drainage costs, and
- social costs, among others.

The intention here is to reemphasize the importance of paying appropriate attention to environmental issues when developing irrigation projects.

14.2 Problems of Flood Control

Perhaps the first problem to be tackled with respect to flood control in Nigeria is the need to establish coordination among the ministries, departments, and other organizations involved with flood control for the proper treatment of the threat of floods which is a nationwide phenomenon.

There is the need to evolve Emergency Preparedness Plans against important flash floods and to organize studies on flood forecasting and management. The above plans and studies should be aimed at the eventual preparation of a national masterplan for flood control.

References

Abalu, G.O. and D'Silva, B. 1980. "Nigeria's Food Situation: Problems and Prospects. Food Policy" : 49 - 60.

Adams, W.M. 1983. "Irrigation as Hazard: Farmers' Responses to the Introduction of Irrigation in Sokoto, Nigeria". In Adams, W.M.; and Grove, A.T. (Eds.). *Irrigation in Tropical Africa: Problems and Problem Solving*. Cambridge African Monographs 3. African Studies Centre, Cambridge.

Adegboye, R.O. 1976. "Land Tenure in Some Parts of West Africa". Department of Agricultural Economics and Extension, University of Ibadan, Ibadan.

Adeniji, F.A. 1985. "Inventory of Irrigation Scope in Nigeria". Federal Ministry of Water Resources and Rural Development, Abuja.

Adeniji, F.A. 1992. "First Update of Inventory of Irrigation Scope in Nigeria: Final Report". Federal Ministry of Water Resources and Rural Development, Abuja.

AERLS 1981. "Proceedings of the 8th National Irrigation Seminar". AERLS, ABU, Zaria.

Adeniyi, F.O. 1973. "Downstream Impact of the Kainji Dam". In Mabogunje, A.L. (Ed.). *Kainji Lake Studies, Vol. 2: Socio Economic Conditions*. NISER, Ibadan.

Agbonifo, P.O. 1974. "Agro-industrialism in the Greater Zaria Area: The Introduction and Effects of Dry-Season Tomato Growing Along the Kubani and Shika River Valley". Unpublished M.Sc. Thesis, Ahmadu Bello University, Zaria, Nigeria.

Agboola, S.A. 1979. *An Agricultural Atlas of Nigeria*. Oxford University Press, Oxford, 248pp.

Agriculture Department. Nigeria. 1918. *Annual Report on the Agricultural Department, Southern Provinces. Agriculture Department. Nigeria. 1922. Annual Report.*

Agriculture Department, Nigeria. 1923. Annual Report.

Agriculture Department, Nigeria. 1926. Annual Report.

Agriculture Department, Nigeria. 1931. Annual Report.

Agriculture Department, Nigeria. 1933. Annual Report.

Agriculture Department, Nigeria. 1934. Annual Report.

Agriculture Department, Nigeria. 1935. Annual Report.

Agriculture Department, Nigeria. 1936. Annual Report.

Agriculture Department, Nigeria. 1940. Annual Report.

Agriculture Department, Nigeria. 1942. Annual Report.

Agriculture Department, Nigeria. 1944. Annual Report.

Agriculture Department, Nigeria. 1945. Annual Report.

Agriculture Department, Nigeria. 1946. Annual Report.

Agriculture Department, Nigeria. 1950. Annual Report.

Agriculture Department, Nigeria. 1954. Annual Report.

Agriculture Department, Northern Region. 1954. Annual Report, 1954-55.

Agriculture Station Reports. 1947. Northern Nigeria Agriculture Station Reports, 1938-1946. National Archives, Kaduna.

Agriculture Station Reports. 1960. *Northern Nigeria Agriculture Station Reports, 1941-1960.* National Archives, Kaduna.

Agriculture Station Reports. 1965. *Northern Nigeria Agriculture Station Reports, 1954-1964,* Ilorin. National Archives, Kaduna.

Akintola, F.O. 1979. "Land Use and Floods in Urban Areas"s. A Keynote Address at the 27th Session of the Nigerian Hydrological Technical Committee held in Ibadan, Oyo State, 30th March, 1979.

Akintola, J. O.1986. *Rainfall Distribution in Nigeria 1892 - 1983.* Impact Publishers (Nig.) Ltd; Ibadan,380 pp.

Aladejana, K. 1982. "Guidelines on National Policy on Agricultural Land Resources. In FDALR (Eds.). *Efficient Use of Nigerian Land Resources.* Federal Ministry of Agriculture and Natural Resources, Abuja: 98-110.

Aminu-Kano, M. 1994. "Sustainable Water Resources Development in the Komadugu-Yobe Basin". Paper Presented at the National Symposium on Water Resources and Rural Development, Kano State House of Assembly, Kano, 5th-10th June, 1994.

Apeldoorn, G. J. van. 1981. *Perspectives on Drought and Famine in Nigeria.* George Allen and Unwin, London.

Awogbade, M.O. 1994. Fadama Resource Use and the Pastoral Question. In Kolawole, A; Scoones, I.; Awogbade, M. O. and Voh, J.P. (Eds.). *Strategies for the Sustainable Use of Fadama Lands in Northern Nigeria.* Centre for Social and Economic Research, Ahmadu Bello

University, Zaria: 97-102.

Awoyemi, O. 1982. "Land Use Act 1978: Its Implications for Agricultural Development in Nigeria". In FDALR (Eds.). Efficient Use of Nigerian Land Resources. Federal Ministry of Agriculture and Natural Resources, Abuja: 74-77.

Balogun, E. E. and Salami, A.T. 1994. "Global Climate Change and Water Resources Potentials in Nigeria". Paper Presented at the National Symposium on Water Resources and Rural Development, Kano State House of Assembly, Kano, 5th-10th June.

Ben-Musa, S. 1992. Operational Problems of Kano River Project Phase I Before and After Commercialisation". Paper Presented at the 1st Annual National Water Resources Forum, National Water Resources Institute, Kaduna, 26th-28th May, 1992.

Brown, J.D. 1928. *Annual Report, Sokoto Irrigation Farm*. 1927. 7th Annual Bulletin, Department of Agriculture, Nigeria: 187-202.

Brown, R. 1982. "Evaluation of Petrol, Diesel and Solar Pumps". KNARDA, Kano.

CBDA. 1984. "Know About Chad Basin Development Authority". CBRDA, Maiduguri.

Chapman, N.P. 1983. "The economic Viability of Improved Irrigation Methods in Northern Zone" BSADP. Paper Presented at the Fadama Seminar > 5th-6th July, at Azare, Bauchi State. Bauchi State ADP, Bauchi.

Chapman, N.P. 1984. "Economic Returns to Pump Users in Northern Zone" BSADP. Paper Presented at the Second Fadama Seminar, 6th-8th March, at Azare, Bauchi State. Bauchi State ADP, Bauchi.

Crowder, M. 1966. *The Story of Nigeria*. Faber and Faber, London, 416pp.

Daily Sketch. 1980. 2nd September, 1980.

Daily Times. 1978. 6th May, 1978.

ECA. 1980. *Land and Water Resources Survey for Irrigation in Africa: Vol. IIA*. Economic Commission for Africa, Addis Ababa, Ethiopia.

Ejieji, C.J. and Nwa, E.U. 1989. "Profile of Irrigation in Kwara State". Paper Presented at the 13th Annual Conference of the Nigerian Society of Agricultural Engineers. Ahmadu Bello University, Zaria, 26th-29th September, 1989.

Evans, T. and Gunganesharajah, K. 1985. "Lake Chad at Risk if Drought Goes on". *World Water*: 34-35.

Famoriyo, S. 1982. "Factors Impinging Upon Land Resources Management in Nigeria". In FDALR (Eds.). *Efficient Use of Nigerian Land Resources*. Federal Ministry of Agriculture and Natural Resources, Abuja: 289-300.

Faure, F. and Gac, J. 1981. "Will the Sahelian drought end in 1985?" *Nature* 291:475-478.

Fawole, M.O. (Ed.). 1980. "Proceedings of a Symposium on the Sugarcane Industry in Nigeria". The Unilorin Sugar Research Institute, Ilorin.

FAO. 1966. "Agricultural Development in Nigeria 1965-1980". FAO, Rome.

FAO. 1992. "Irrigation Sub-Sector Review, Nigeria". FAO, Rome.

FMANR. 1984. "Information Bulletin on Nigerian Agriculture". Federal Ministry of Agriculture and Natural Resources, Abuja.

FMANR, File No. AGR 137/S.1. 1968. International Commission on Irrigation and Drainage. Federal Ministry of Water Resources and Rural Development, Abuja.

FMWR, File No. FDWR/WR/S.57. 1985. River Niger Basin Authority. Federal Ministry of Agriculture and Natural Resources, Abuja.

Green Revolution National Committee. 1981. "Report of Research Institutes Review Panel 1980/81, Volume 1: Background Papers and Evaluation of Individual Research Institute". Moor Plantation, Ibadan.

Griffith, W.J. 1983. "Irrigation in Nigeria: Madness of Fowl, Marriage with Cat?" In Adams, W.M.; and Grove, A.T. (eds.). Irrigation in Tropical Africa: Problems and Problem-Solving. Cambridge African Monographs 3. African Studies Centre, Cambridge.

Hamilton, R.E. 1969. "Damodar Valley Corporation: India's Experiment with the T.V.A. Model". *Indian J. of Public Administration* 15(1): 86-109.

Hardcastle, J.E. 1959. "The Development of Rice Production and Research in the Federation of Nigeria". *Tropical Agriculture*, Trinidad 36(2): 79-95.

Haskoning Engineering Consultants, Nigeria. 1978. "Tiga Dam Evaluation Report". Hadejia-Jama'are River Basin and Rural Development Authority, Kano.

Havinden, M.A. 1975. "The History of Crop Cultivation in West Africa: A

KMAG, File No. 11019. Vol. I. 1924. "Preliminary Report on Irrigation Part II". National Archives, Kaduna.

KMAG, File No. 30867. 1925. "Shalla Project: Report and Estimate", by Col. M.R. Collins, Irrigation Engineer. National Archives, Kaduna.

KMAG, File No. 30867. 1926. "Report on Irrigation in Sokoto Province", by A.L.V. Long. National Archives, Kaduna.

KMAG, File No. 82/1924/56. 1926. "Irrigation: Progress Report of Work in Sokoto". National Archives, Kaduna.

KMAG, File No. 1060 Vol. II. 1932. "Kware Irrigation Scheme". National Archives, Kaduna.

KMAG, File No. 1060 Vol. III. 1934. "General Report on Kware Irrigation Farm and Scheme", by J.H. Palmer, Superintendent of Agriculture. National Archives, Kaduna.

KMAG, File No. 26902. 1937. "Irrigation in Northern Provinces". National Archives, Kaduna.

KMAG, File No. 2702. 1947. "Wuya-Edozhigi Irrigation Scheme". National Archives, Kaduna.

KMAG, File No. 42627. 1947. "Egwa River Dam". National Archives, Kaduna.

KMAG, File No. 47678. 1949. "The Wuya-Edozhigi Irrigation Project". National Archives, Kaduna.

KMAG, File No. 2700A. 1951. "Rice Project, Kwa Falls Estate, Calabar: Preliminary Report on Flood Protection". National Archives, Kaduna.

KMAG, File No. 47678. 1951. "Wuya-Edozhigi Irrigation Project". National Archives, Kaduna.

KMAG, File No. 26902. 1953. "Outline of Irrigation Policy". National Archives, Kaduna.

KMAG, File No. 4637/S.I. 1953. "Irrigation and Drainage in Nigeria". National Archives, Kaduna.

KMAG, File No. 72390. 1953. "Tiga Rapids-Kogin Kano". National Archives, Kaduna.

KMAG, File No. 2705C. 1955. "Irrigation Scheme, Badeggi". National Archives, Kaduna.

KMAG, File No. 49845. 1956. The Native Authority Law No. 4 of 1954. Date of Commencement 1st May, 1956. National Archives, Kaduna.

KMAG, File No. IRR 104. 1958. "Kware Irrigation Scheme, Sokoto". National Archives, Kaduna.

KMAG, File No. IRR. 104: 1959. "Wurno Irrigation Scheme". National Archives, Kaduna.

LCBC 1973. Lake Chad Basin Commission: Convention and Statute Relating to the Development of the Chad Basin. Lake Chad Basin Commission N'Djamena, Chad.

LCBC 1990. "Report of the Fourth Conference of Ministers of Environment to Save Lake Chad". Lake Chad Basin Commission, N'Djamena, Chad.

Linsley, R.K. and Franzini, J.B. 1964. *Water Resources Engineering*. New York: McGraw-Hill.

Lot, E.Y. 1982. "Land Use Act 1978: Its implications for Agricultural Development in Nigeria". in FDALR (Eds.) *Efficient Use of Nigerian Land Resources*. Federal Ministry of Agricultural and Natural Resources, Abuja: 83-86.

Makanjuola, F.J. and Nwa, E.U. 1985. "Effect of Irrigation Water Management on the Yield of Sugar Cane at the Bacita Estate, Nigeria". *Proceedings, 5th Afro-Asian Regional Conference of the International Commission of Irrigation and Drainage*, 25th-30th August, 1985, Townsville, Australia: 138-145.

Michael, A.M. 1978. *Irrigation, Theory and Practice*. New Delhi, India: Vikas Publishing House, PVT, Ltd, 801p.

Mijindadi, N.B; Umar, Q; and Tyem, M.N. 1993. "ADP Experiences with Fadama Development: Achievements, Problems and Prospects". in Nwa, E.U; and Pradhan, P. (eds.). *Irrigation Research Priorities for Nigeria*. International Irrigation Management Institute, Colombo, Sri Lanka.

Ministry of Agriculture, Northern Nigeria. 1960. *Annual Report*.

Ministry of Agriculture, Northern Nigeria. 1961. *Annual Report*.

Ministry of Agriculture, Northern Nigeria. 1965. *Annual Report*, 1965/66.

National Archives, Enugu. 1945. File No. CSE1/85/5261, National Archives, Enugu.

NBTE 1985. *The Directory of Technical and Technological Institutions in Nigeria*. National Board for Technical Education, Kaduna.

NEDECO 1959. *River Studies and Recommendations and Improvement of Niger and Benue*. North-Holland Publishing Company, Amsterdam, The Netherlands, and NEPA, Lagos.

NEDECO 1970. "Gongola River Project Situated in the North-Eastern State of the Federal Republic of Nigeria - Reconnaissance Report". NEPA, Lagos.

NHTC 1974. "Report of the Hydrological Technical Sub-Committee on Drought". Federal Ministry of Water Resources and Rural Development, Abuja.

Nigeria 1963. *National Development Plan 1962-68.* The Federal Ministry of Economic Development, Lagos.

Nigeria 1975. *Third National Development Plan 1975-80. Vol. I. Main Report.* Federal Ministry of Economic Development, Lagos.

Nigeria 1978. "Land Use Decree No. 6 of 1978". *Official Gazette Vol. 65 No. 14.* Lagos: Government Printer.

Nigeria 1987. *River Basins Development Authorities Decree No. 35 of 1987.* Lagos: Government Printer.

Nigeria 1993. "Water Resources Decree 1993". *Official Gazette Extraordinary No. 27 Vol. 80.* Lagos: Government Printer.

NSC n.d. The Nigerian Sugar Company Limited. Bacita, Kwara State.

Nwa, E.U. 1975. "Economics of a Small-Scale Trickle Irrigation System". *Proceedings of the National Irrigation and Agricultural Engineering Seminar*, August 1975, Kano. Federal Ministry of Water Resources and Rural Development, Abuja.

Nwa, E.U. 1980. Report of a Questionnaire.

Nwa, E.U. 1981a. "An Evaluation of Small Pump and Shaduf Systems of Irrigation in Northern Nigeria". Samaru J. Agric. Res. 1(12): 191-201.

Nwa, E.U. 1981b. "Using the Bamboo (Bambus vulgaris) as Field Drainage Material". Trans. American Society of Agricultural Engineers 24(4): 974-976.

Nwa, E. U. 1981c. "Wind Characteristics for Pumping Water in Zaria, Nigeria". *Proceedings of the 8th National Irrigation Seminar.* NAERLS, Zaria: 100-113.

Nwa, E.U. 1982. "Groundwater Problems of the Kano River Irrigation Project of Nigeria". *Proceedings of the Fourth Afro-Asian Regional Conference of the International Commission on Irrigation and Drainage*, Lagos, Nigeria, Vol. II, Publ. 10: 123-137.

Nwa, E.U. 1983. "Role of Irrigation Research in Agricultural Development in Nigeria". *Water World* 5(3): 17-20.

Nwa, E.U. 1987. "Irrigation and Nigeria". An Inaugural Lecture, University of Ilorin, 40pp.

Nwa, E.U. 1993a. "Irrigation in Competi·ion for Water: The Case of the Lake Chad Basin". *Fifteenth Congress of the International Commission on Irrigation and Drainage*, The Hague. Special Session R. 9: 115-127.

Nwa, E.U. 1993b. "The status of irrigation research in Nigeria". in Nwa, E.U; and Pradhan, P. (Eds.). *Irrigation Research Priorities for Nigeria.* International Irrigation Management Institute, Colombo, Sri Lanka: 11-18.

Nwa, E.U. 1994a. "Appropriate Irrigation and Drainage Practices to Support the Index Crops for the Various Ecological Zones of Nigeria". Paper Presented at the National Symposium on Water Resources and Rural Development, Kano State House of Assembly, 5th-11th June, 1994. Federal Ministry of Water Resources and Rural Development, Abuja.

Nwa, E.U. 1994b. "Participatory Irrigation Management in Public Irrigation Schemes: The Case of Kano River Irrigation Project, Nigeria". Paper presented at the Consultation on Irrigation Extension, FAO, Accra, Ghana, 5th-9th December, 1994.

Nwa, E.U. 1994c. "Appropriate Management of Irrigated Agriculture in Nigeria". Paper Presented at the River Basin Irrigation Planning Study Workshop: An African Development Bank Study Orientation Workshop at the National Water Resources Institute, Kaduna, 18th - 21st October, 1994.

Nwa, E.U. and Martins, B. 1982. "Irrigation Development in Nigeria". Proceedings of the Fourth Afro-Asian Regional Conference of the International Commission of Irrigation and Drainage, Lagos, Nigeria, Vol. I, Publ. 01: 1-18.

NWRI 1984. The National Water Resources Institute, Kaduna: *What You Need to Know.* National Water Resources Institute, Kaduna.

NWRMP 1995. T*he Study on the National Water Resources Master Plan.* Sector Report, Volume Two, Part 1. Federal Ministry of Water Resources and Rural Development, Abuja.

Odihi, J.O. 1994. "Fadama Resource Sustainability: Lessons from the Jere Rice Bowl Experience, Borno State". In Kolawole, A; Scoones, I; Awogbade, M.O.; and Voh, J.P. (Eds.). *Strategies for the Sustainable Use of Fadama Lands in Northern Nigeria.* Ahmadu Bello University, Zaria.

Olofin, E.A. 1993. "Sustainable Irrigation Option for Nigeria: The Large or the Small". In Nwa, E.U. and Pradhan, P. (eds.). *Irrigation Research Priorities for Nigeria*. International Irrigation Management Institute, Colombo, Sri Lanka.

Orewa, S. I. 1978. "The Economics of Dry Season Tomato Production Under the Cadbury Tomato Scheme in Zaria Area". Unpublished M.Sc. Thesis, Ahmadu Bello University, Zaria, Nigeria.

Orewa, S.I. 1981. "Some Technical and Management Problems in Small-holder Irrigation Projects: A case study from Zaria area". *Proceeding of the 8th National Irrigation Seminar*: 136-145.

Oyebande, L. and Nwa, E.U. 1980. "Effects of Drought and Water Projects on the Water Balance and Flow Regime in the Upper Hadejia Basin, Nigeria. *Proceedings of the Helsinki Symposium*, IAHS, Publ. No. 130: 215-220.

Palmer-Jones, R.W. 1981. "How Not to Learn from Pilot Irrigation Projects": The Nigerian Experience. *Water Supply and Management* 5: 81-105.

Quinn, J.G. 1971. "Factors Affecting the Production of Tomatoes for the Processing Industry During Different Seasons in the Northern States of Nigeria". Unpublished M.Sc. Thesis, Ahmadu Bello University, Zaria, Nigeria.

Pedder, J. B. 1937, *Windmills for pumping in Nigeria*. Government Printer, Lagos.

Rae, C.J. 1944. "Report on Possible Agricultural Development by Means of Irrigation and Drainage in the Colony and Protectorate of Nigeria". KMAG, File No. 26902. National Archives, Kaduna.

Raza, M.R. and Abdullahi, Y.A. 1979. "Irrigation Schemes and Agrarian Structure: A re-specification of the Impact of the Structure of Landholding and Land use Patterns on Agricultural Development". Proceedings of the 6th National Irrigation Seminar. AERLS, Ahmadu Bello University, Zaria: 216-227.

Rydzewski, R.J. 1972. "Prospects for Irrigation in West Africa". An International Seminar. The Conference Centre, University of Ibadan, Ibadan, Nigeria.

Sadiq, M. 1994. "Issues of Dam Safety and Reservoir Operation in Nigeria". Paper Presented at the National Symposium on Water Resources and Rural Development, Kano State House of Assembly, 5th-11th June, 1994.

Sagua, V.O. 1977. "Flood Control of the River Niger at Kainji Dam, Nigeria, and its Use During Drought Conditions, the Period 1970-1976". *Proceedings of the Conference on the Aftermath of Drought*, Kano, April, 1977: 127-140.

Saidu, M.M. 1994. "Land and Water Rights in Fadama Lands of Northern Nigeria. In Kolawole, A; Scoones, I; Awogbade, M.O; and Voh, J.P. (Eds.). *Strategies for the Sustainable Use of Fadama Lands in Northern Nigeria.* Centre for Social and Economic Research, Ahmadu Bello University, Zaria: 91-95.

Schultz International Limited 1975. "Hadejia River Basin Study, Volume A: Development Potential, and Volume F: Land Use and Water Rights Study". Federal Ministry of Water Resources and Rural Development, Abuja.

Singh, A. and Maurya, P.R. 1979. "Salinity Status and Waterlogging in Irrigated Agriculture of Northern Nigeria". *Proceedings of the 6th National Irrigation Seminar*, Zaria: 146-163.

Smith, N. 1975. *Man and Water: A History of Hydro-Technology*. Richard Clay Ltd., Stel, London.

SOKPROF 1950 File No. 6613. "Dams for Cattle Watering". National Archives. Kaduna.

SRRBDA 1982. "Report of the General Manager". 9th National Irrigation Seminar, 13th-17th September, 1982. Sokoto. Sokoto-Rima River Basin and Rural Development Authority, Sokoto.

SRRBDA 1983. Commissioning of the Bakolori Irrigation Project, 9th April, 1993. Sokoto-Rima River Basin and Rural Development Authority, Sokoto.

Stride, G.T. and Ifeka, C. 1971. *Peoples and Empires of the West Africa (West Africa in History, 1000-1800)*. Africana Publishing Corporation, New York.

Sunday Sketch 1980. 14 September, 1980.

Tarhule, A. and Woo, M. 1997. "Towards an Interpretation of Historical Droughts in Northern Nigeria". *Climatic Change* 37: 601-616.

Thambyahpillay, G.R. 1983. "The Palaco-hydroclimatology of Lake Chad. *Annals of Borno, Vol. I*: 105-145.

Turner, B. 1977. "The Fadama Lands of Central Northern Nigeria: Their Classification, Spatial Variation, Present and Potential Use". Ph.D

Thesis, University of London. Volume I, 297pp.

UNDP/FAO. 1969. "Sokoto-Rima Valley: Pre-feasibility Studies of Soil and Water Resources of the Sokoto Valley. Nigeria". FAO, Rome.

USDI 1968. "Reconnaissance Study: Land and Water Resources of Lake Chad Basin". U.S. Bureau of Reclamation. Denver, Colorado.

Wood, A. D. 1976. "Water Lifters and Pumps for the Developing World". Master of Science Thesis. Colorado State University, Fort Collins, Colorado, U.S.A.

World Bank 1977. "Appraisal of Lafia Agricultural Development Project. in Nigeria". World Bank, Washington D.C; U.S.A.

World Bank 1982. "Staff Appraisal Report, Nigeria: Sokoto Agricultural Development Project". World Bank, Washington D.C., U.S.A.

World Bank 1992. "Staff Appraisal Report", Federal Republic of Nigeria: National Fadama Development Project. World Bank, Washington D.C., U.S.A.

WRECA. n.d. "The Services of WRECA". Water Resources and Engineering Construction Agency, Kano.

Zayyad, H.R. 1992. "Commercialization of Water Resources Agencies in Nigeria". *Proceedings of the National Water Rehabilitation Project Launch Workshop* at Sheraton Hotel & Towers. Abuja, 22nd - 27th November, : 267-278.

Zimmerman, J.D. 1966. *Irrigation.* New York : John Wiley & Sons, Inc., 516pp.

APPENDICES

Appendix A.1

Table A. 1: Existing small irrigation dams, 1992

General			Reservoir			Dam					
State Dam	Agency	River	Area ha	Capacity MCM Total	Capacity MCM Active	Type	Height m	Length m	Volume $10^3 m^3$	Year of completion	Purpose irrigation plus
1. Abia	AIRERDA	Igwu	-	0.4	0.3	earth	5.0	-	-	-	S
(i) Igbere											
(ii) Umuopara	MANR	Aba	-	-	-	"	-	220	-	-	
2. Bauchi											
(i) Cham	UBRERDA	Cham	180	8.0	6.5	"	8.0	1,300	-	1992	S
(ii) Galala	URARBRDA	Galala	110	23.0	20.0	"	12.0	400	-	1982	
3. Benue											
(i) Naka	LBRERDA	Ana	150	2.5	2.0	"	7.0	500	-	1986	S
(ii) Umogidi	LBRERDA	Ogabakpa	100	2.5	1.8	"	5.0	310	-	1985	S
4. Borno											
(i) Alau	CBRDA	Ngadda	5,000	112.4	106.0	"	9.5	344	-	1992	S
5. Edo											
(i) Okweghi	MANR	Okweghi	-	0.8	0.6	"	0.1	12	-	1992	S
6. Jigawa											
(i) Ibrahim											
(ii) Adamu	WRECA	Gari	260	8.0	7.4	"	91	183	-	1974	S1
(iii) Wawarde	WRECA	Dunduru									
		Wawarde	530	12.3	9.7	"	100	2,780	-	1992	S
Hadejia	HURBRDA	Hadejia	2,000	-	14.0	concrete	-	-	-	1992	

Note refers to purpose

1. blank - irrigation is sole purpose
2. S - irrigation main purpose, water supply secondary purpose
3. S1 - irrigation secondary purpose, water supply main purpose

Table A. 1 cont'd.

General		Reservoir				Dam					
State Dam	Agency	River	Area ha	Capacity MCM Total	Capacity MCM Active	Type	Height m	Length m	Volume 10^3 m^3	Year of completion	Purpose irrigation plus
7. Katsina											
i) Jibiya	SRRBRDA	Gada	2,600	142.0	121.0	earth	21.1	3,680	2.7	1990	S
ii) Zobe	”	Karaduwa	4,500	177.0	170.0	”	18.9	2,750	3.2	1983	S
8. Kogi											
i) Omi	LNRBRDA	Kampe	2,570	250.0	220.0	”	43.0	1,988	3.8	UC	S
9. Kwara											
i) Ero	SWA	Ero	-	20.0	18.5	”	22.0	662	-	1987	SI
10. Niger											
i) Kainji	NEPA	Niger	127,000	12,000	9,140.0	concrete	65.5	5,300	6.1	1968	PI
ii) Kubli	UNRBRDA	Swashi	940	75.0	62.0	”	23.0	128	-	1992	
iii) Swashi	”	Swashi	-	16.0	12.6	earth	17.6	800	0.3	-	
iv) Kontagora (2)	”	Kontagora	3,900	340.0	200.0	”	32.0	1,400	1.4	UC	
v) Guzan	MANR	Yiko	-	20.0	18.0	”	-	-	-	-	
11 Ogun											
i) Oyan	OORBRDA	Oyan	4,000	270.0	254.0	”	30.4	1,044	1.1	1983	SP
12. Oyo											
i) Igbojaiye	OORBRDA	Oye	-	5.6	5.2	earth	18.0	-	1.4	UC	S
ii) Ikere Gorge	OORBRDA	Ogun	5,300	690.0	565.0		47.5	660	1.1	1983	SP
13. Plateau											
i) Bokkos (1)	LBRBRDA	Maber	-	5.6	5.2	”	15.0	340	0.5	-	S
ii) Doma	LBRBRDA	Ohna	220	37.5	30.0	”	15.7	520	3.5	1988	S
14. Sokoto											
i) Gusau	SWA	Sokoto	-	3.0	2.5	”	22.0	800	-	-	SI
ii) Bakolori	SRRBRDA	Sokoto	8,000	450.0	403.0	”	48.0	5,500	-	1978	SP
iii) Goronyo	”	Rima	20,000	942.0	933.0	”	20.0	5,285	4.5	1984	S

Appendix A.2

Table A. 2: Existing large and medium dams for irrigation, 1992

General				Reservoir							Dam
State Dam		Agency	River	Area ha	Capacity MCM Total	Capacity MCM Active	Type	Height m	Length m	Volume 10^3 m^3	Year of completion	Purpose irrigation plus
1.	Adamawa											
i)	Kiri	UBRBRDA	Gongola	11,000	651.0	350.0	earth	37.0	1,300	-	1982	
2.	Bauchi											
i)	Balanga	MANR	Balanga	1,1000	73.0	63.0	"	41.0	180	04	1987	
ii)	Dadin Kowa	UBRBRDA	Gongola	30,000	2,8555.0	1,770.0	"	42.0	520	11	1988	SP
iii)	Waya	"	Waya	450	21.0	17.0	"	23.0	400	-	UC	S
iv)	Kafin Zaki	HURBRDA	Bunga	23,500	2,700.0	2,500.0	"	40.0	11,000	19.0	UC	
3.	Borno											
i)	Biu	OBRDA	Nidivana	300	11.9	9.6	earth	-	-	-	-	S
4.	Cross River											
i)	Obudu	CRBRDA	Aya	50	2.5	2.0	"	15	-	-	UC	SI
5.	Kaduna											
i)	Bagoma	SWA	Kusheriki	-	5.5	5.0	"	17.0	2.225	-	-	SI
ii)	Kangimi	SWA	Kangimi	600	74.1	59.3	"	19.2	1.524	-	1975	SI
6	Kano											
i)	Gari	MANR	Gari	3,320	214.0	203.0	"	22.0	6.055	3.1	1980	S
ii)	Challawa Gorge	HURBRDA	Challawa	10.000	930.0	900.0	"	38.0	7.804	6.2	1992	S
iii)	Guzuguzu	MANR	Guzuguzu	640	24.6	21.5	"	17.4	2.090	0.7	1992	
iv)	Magaga	MANR	Magaga	370	19.7	17.2	"	19.4	2,550	0.7	1980	S
vi)	Watari	KNARDA	Watari	1,960	104.5	21.5	"	19.8	3.658	1.9	1980	S
vii)	Tudun Wada	MANR	Waina	350	20.8	16.6	"	21.0	2.478	-	1977	
viii)	Tiga	HURBRDA	Kano	17,800	1.9680	1,845.0	"	47.2	5,791	10.8	1975	S
ix)	Bagauda	MANR	Baguada	376	22.1	27.1	"	20.7	2.134	0.7	1970	
ix)	Ruwan Kanya	HURBRDA	Kanya	-	-	-	"	22.0	3.658	-	1976	S
x)	Kafin Chiri	MANR	Jatau	840	31.1	31.1	"	16.0	5,405	6.0	1977	

<u>Note refer to purpose</u>

1 Blank - irrigation is sole purpose
2. S - irrigation main purpose, water supply secondary purpose
3. SI.PI - irrigation secondary purpose, water supply or power main purpose
4. SPs - irrigation main purpose, water supply and/or power secondary purpose

Table A. 2 contd.

General			Reservoir			Dam					
State Dam	Agency	River	Area ha	Capacity MCM Total	Capacity MCM Active	Type	Height m	Length m	Volume $10^3 m^3$	Year of comple-tion	Purpose irriga-tion plus
7. Kaduna											
i) Birnin Gwari	SWA	Kusheriki	200	4.0	3.2	earth	14.3	1,657	-	1972	SI
ii) Kerewa	UNRBRDA	Tubo	-	3.5	2.8	"	7.0	600	-	1988	S
iii) Kargi	MANR	Kargi	3,500	0.5	0.3	"	2.9	226	-	1987	
iv) Ikara	SWA	Kariga	-	3.4	3.0	"	10.5	710	-	-	SI
v) Matari	MANR	-	3,600	12.6	8.2	"	7.5	550	-	1987	
vi) Paki	MANR	Paki	40,000	0.3	0.3	"	7.0	690	-	-	
vii) Wandon	MANR	-	"	0.2	0.2	"	6.0	250	-	'988	
viii) Bomo	MANR	Bomo	27	0.25	-	"	-	387	7.6	1961	
8. Kano											
i) Tomas	MANR	Tomas	1,500	60.3	56.6	"	13.7	3,353	-	1976	
ii) Jakara	WRECA	Jakara	1,600	65.4	54.5	"	14.3	2,438	-	1976	
iii) Marachi	WRECA	Marachi	210	6.8	5.8	"	11.5	540	-	1980	
iv) Pada	HJRBDA	Pada	410	12.0	10.5	"	14.2	2.151	-	1980	SI
v) Kango	WRECA	Kango	260	8.7	7.9	"	14.5	1.121	-	UC	SI
9. Katsina											
i) Tubo	SWA	Tubo	1,150	35.5	25.1	"	8.4	780	-	1983	SI
ii) Mairuwa	SWA	Sokoto	160	5.3	5.3	"	12.0	450	-	1970	SI
10. Kebbi											
i) Ililo	MANR	Ilogourou	500	-	-	"	5.0	420	-	1990	S
11. Niger											
i) Rijau	UNRBRDA	Butulu	-	1.3	1.0	"	7.0	410	-	1990	
ii) Nasko	UNRBRDA	Shadagulbi	60	2.6	2.0	"	12.0	645	-	1990	S
iii) Tungan Kawo	UNRBRDA	Ubandawaki	400	22.0	21.0	"	11.8	3,300	-	1888	S
12. Ogun											
i) Lekan Are	OORBRDA	Ogun	-	0.6	0.6	"	10.0	1,600	-	1982	S
ii) Oke Odan	"	Oke Odan	-	7.5	7.2	"	-	-	-	UC	S

Table A.2 contd.

General			Reservoir			Dam					
State Dam	Agency	River	Area ha	Capacity MCM Total	Active	Type	Height m	Length m	Volume 10^3 m^3	Year of completion	Purpose Irrigation plus
13. Ondo											
i) Apariko	MANR	Apariko	-	4.7	4.0	zone earth	0.2	13	-	-	S
ii) Owena	SWA	Owena		-	9.0	earth	9.0	-	-	1982	SI
14. Oyo											
i) Ofiki (A)	OORBRDA	Ofiki	-	1.3	1.1	"	12.6	580	-	1983	S
ii) Ofiki (B)	OORBRDA	Ofiki	-	0.6	0.5	"	12.3	520	-	1981	S
iii) Igboho	MANR	Ogun	-	1.2	0.9	earth	10.0	-	-	1988	S
iv) Sepeteri (A)	OORBRDA	Owutu	-	2.6	2.4	"	13.6	685	-	1984	S
v) Sepeteri (B)	"	Agbado Osai inm	1.9	1.3	-	"	13.5	-	-	1989	S
vi) IITA	IITA	Awba	7.0	1.5	-	"	12.2	720	-		S
15. Plateau											
i) Tolle Maribe	SWA	Rafin Sanyi	10	-	-	concrete	7.6	374	-	1961	SI
ii) Shedam (I)	UBRBRDA	Tunkua	240	4.5	3.4	earth	13.0	1,200	-	1984	SI
iii) Kwall	LBRBRDA	Ouree	35.000	0.2	-	"	6.0	180	-	UC	S
16. Sokoto											
i) Wurno	SRRBRDA	Rima	-	25.0	20.0	"	8.0	800	-	1960	

Appendix A.3

Table A. 3 Small-scale Irrigation schemes

State Schemes	Owner if not State	Proposed area (ha)	Area irrigated by Dec. 1992 (or by Dec. 1980) ha	Year of 1st irrigation	Source of water	Crops
1. Abia						
i) Bende	AIREBRDA	300	150		Intake of Igwu river	rice
ii) Ndojinguwa Abam		96	0		Pumping	rice
iii) Ozu Abam		150	50		Pumping from Igwu river	rice
iv) Owutu Eddah		480	280		Pumping from Asu river	rice
v) Idim Abam		150	50		Pumping from Igwu river	rice
vi) Amaeke Abam		23	23			rice
vii) Ofiavu Ozuitem		185	0			rice
viii) Akaeze		200	100		Pumping from Asu river	rice
ix) Umuopara		200	100		dam across Aba river	rice
x) Umuhu		400	200		Pumping from Igwu river	rice
xi) Igwu	AIREERDA				Pumping from Igwu river	
a. Ndiebe		200	100			rice
b. Ndiojo Ugwo		100	0			rice
c.) Ohafia		300	20			rice
xii) Upiti		60	10			rice
2. Adamawa						
i) Talllum	UBRBRDA	400	160		Pumping from Gongola river	
ii) Lake Geriyo Farm	UBRBRDA	50	(24)		Pumping from Lake Geriyo	vegetables
iii) Mayo-Bani		40			Pumping from river Yedseram	rice, maize, vegetables
iv) Dasin		25				rice, maize, sugar-cane
v) Jimeta Garden		18				Vegetable, maize
vi) Diladum		6				Vegetable,rice, sugar-cane
vii) Loko		4				Vegetable, maize
viii) Garkida - Garden		4				Vegetable, maize

Table A. 3 cont'd.

State	Schemes	Owner if not State	Proposed area (ha)	Area irrigated by Dec. 1992 (or by Dec. 1980)	Year of 1st irrigation	Source of water	Crops
3.	Akwa Ibom						
i)	Abak	CRRDA	90	60	1980	Pumping from Abak river	maize, vegetables
ii)	Nung Obong		200	100		Pumping from Ubum river	cowpea, ginger
iii)	Ikot Inyang (400ha), Ikot Ebidang (100ha), Ikot Obong, Itu (200ba): Ikpa Uruan (60ha): all of these were proposed and some development made but they were all abandoned before 1990					Pumping from Ubum river	maize, vegetables
4.	Anambra						
i)	Enuguabor Ufomia						
	a) Ozi, Enuguabor		260	(180)	1979	Pumping from Ozi stream	rice
	b) Omogho		261	(0)			
	c) Akpago		100	(30)	1983		rice
	d) Aguabosi, Ugwuoba		100	(0)			
ii)	Ogboji		130	30	1965	Pumping from Ubo river	rice
iii)	Ikem Nando	AIBRDA	10	10		Pumping from Ezu river	rice
iv)	Odekpe		120	1		Pumping from Utembo/ Ntiya stream	
v)	Ifite Ogwari		120	30			
5.	Bauchi						
i)	Lake Diya		250	30	1978	Pumping from Lake Diya	vegetables
ii)	Lake Maladumba		250	20	1979	Pumping from Lake Malamduba	vegetables
iii)	Cham		250	150		Cha dam across Cham river	maize, vegetables

Table A. 3 cont'd.

State Schemes	Owner if not State	Proposed area (ha)	Area irrigated by Dec. 1992 (or by Dec. 1980)	Year of 1st irrigation	Source of water	Crops
6. Benue						
i) Odugbo		150	(15)	1972	Pumping from Odugbo-Idi river	maize, vegetables, sugarcane
ii) Aketa		80	10	1978	Pumping from Katsina-Ala	vegetables
iii) Agbene		53	(25)	1978	Pumping from Okpokwu river	rice, maize,
iv) Aokpe		6	(6)	1978	Pumping from Umabe stream	vegetables, maize
v) Mu River		100	60	1979	Pumping from Mu river	vegetables, rice,
vi) Allan		50	10	1978	Pumping from Okpo Kira river	rice, maize, vegetables
vii) Mbayion (Gboko)		12	2	1981	Pumping from Konshisha	river, vegetables
7. Borno						
i) Gamboru	CBRDA	400	150	1964	Pumping from Ebeji river	wheat
ii) Vegfru	VEGFRU	300	300	-	Pumping from Gongola river	vegetables
iii) Pumbum*		200	0	-	water from spring	
iv) Arege		160	(120)	1975	Pumping from Komadugu-Yobe river	rice
v) Damasak		100	30	1985	Pumping from Yobe river	
vi) Lake Alau		30	(30)	1974	Pumping from Lake Alau	vegetables
vii) Munguno		20	10	1977	borehole	wheat, vegetables
viii) Konduga		20	(16)	1979	Pumping from Ngadda river	vegetables
ix) Buluma maduri		12	(12)	1974	artesian well	vegetables
x) Gora	BORNO	50	0	1984	Pumping from Gongola river	

*About 3 ha of this project was developed and abandoned because of constant harassment by elephants which damaged crops and destroyed the irrigation structrures.

Table A. 3 cont'd.

State Schemes	Owner if not State	Proposed area (ha)	Area irrigated by Dec. 1992 (or by Dec. 1980)	Year of 1st irrigation	Source of water	Crops
xi) North Chad after operating this project for a while, it was abandoned because Lake Chad receded away from the project site		40	0	1966	Pumping from Lake Chad	
8. Cross River						
i) Ogoja	CRERDA	250	10	1989	Pumping from Aya river	maize, vegetables
ii) Owakande	"	30	16	1990	water from Obubra stream	maize, vegetables
9. Edo						
i) Ukhun Erah	BORBRDA	250	0		Pumping from Erah river	rice, vegetables
10. Ekiti						
i) Osun Ekiti	BOEBRDA	420	0		intake on Awo river	
11. Enugu						
i) Iboko		300	30		intake on Aboine river	rice
ii) Itemikwo (300ha), and Noyo (100ha) were proposed and some development undertaken but they were abandoned before 1990						
12. Imo						
i) Ihite-Uboma		310	200		weir	rice
13. Jigawa						
i) Hantsu		320	0		Pumping from Hadejia river	rice, vegetables

Table A. 3 cont'd.

State Schemes	Owner if not State	Proposed area (ha)	Area irrigated by Dec. 1992 (or by Dec. 1980)	Year of 1st irrigation	Source of water	Crops
ii) Yamidi		220	(220)	1965	Pumping from Hadejia river	rice, vegetables
iii) Warwade		200	0		dam across Warwade river	rice, vegetables
iv) Gunuar Kukaf	HUBRDA	90	60(1973)	1960	Pumping from Hadejia river	wheat
v) Mai-alkama	"	80	70(1973)	1960	Pumping from Hadejia river	wheat
vi) Arawa	"	80	73(1973)	1960	Pumping from Hadejia river	wheat
vii) Kafin Gana	JIGIWA	60	30		Pumping from Kafin Gana lake	wheat, vegetables
viii) Agubu	HUBRDA	30			Pumping from Hadejia river	wheat
ix) Sunamu	"	30			Pumping from Hadejia river	"
x) Tashina	"	25	25(1973)		Pumping from Hadejia river	"
xi) Sule Tankarkarn (Yandamu)		20	10(1974)	1974	borehole	vegetables
xii) Buluwa		40			borehole	wheat
xiii) Lallashi		20			borehole	orchard
xiv) Jigawa		20	10		borehole	orchard
xv) Dambo		200	0		dam across Gari river	wheat, rice, vegetables
14. Kaduna						
i) Birnin Gwari		430	50	1976	dam across Kusheriki river	wheat, maize, vegetables
ii) Kogun		400	50			
iii) Gurara/Sakin Fawa		300	0		Pumping from Sarkin Fawa & Gurara river	wheat, maize, vegetables
iv) Kidandan		200	40		dam across Tubo river	
v) Kirami		200	0		Karami river + shallow tubewells	maize, wheat, vegetables
vi) Kagoro		120	73	1976	Pumping from Kogun river	rice, vegetables
vii) Shika		100	81	1976	Zaria dam	tomato
viii) Kazuntu		80	65	1972	Pumping from Galma river	vegetables

Table A. 3 cont'd.

State	Schemes	Owner if not State	Proposed area (ha)	Area irrigated by Dec. 1992 (or by Dec. 1980)	Year of 1st irrigation	Source of water	Crops
ix)	Hunkuyi		60	40	1977		
x)	Tabanni		60	4	"		vegetables
xi)	Bagowa		40	17	"		vegetables, maize
xii)	Kuda		24	24	"	Pumping from Kachia river	wheat, vegetables
xiii)	Kaduna Garden		20	20	"	Pumping from Kaduna river	rice, maize, vegetables
xiv)	S. Birni		200	40		intake on Tubo river	
xv)	G. Kurama		200	40		dam across K. Kuri river	
xvi)	Lere		200	40		weir across Karami river	
xvii)	Dawanta		100	40		weir across Galma river	
xviii)	Kerawa	UNRBRDA	100	100	1992	dam across Tubo river	rice, maize, vegetables
xix)	Egunre	LBRBDA	100	100		Pumping from Erubu river	
15.	Kano						
i)	Tundun Wada		360	0		dam across Waina river	wheat, vegetables
ii)	Wala		150	50		Pumping from Hadejia river	wheat, vegetables
iii)	Tsuwa		120	0		Pumping from Hadejia river	wheat, rice, vegetables
iv)	Jakarade		104	102(1975)	1969	Pumping from Gari river	wheat, vegetables
v)	Joda		100			Pumping from Hadejia river	wheat, rice, vegetables
vi)	Balare		40	40		Pumping from Hadejia river	wheat, rice, vegetables
vii)	Bagwai		40			Watari dam	wheat, rice, vegetables
viii)	Dutsen Karya		40			Watari dam	wheat, rice, vegetables

Table A. 3 cont'd.

State	Schemes	Owner if not State	Proposed area (ha)	Area irrigated by Dec. 1992 (or by Dec. 1980)	Year of 1st irrigation	Source of water	Crops
ix)	Idris Gawo		40			Idris Gawo dam	wheat, rice, vegetables, maize
x)	Kunchi		40			Gari dam	wheat, rice, vegetables
xi)	Magaga		40			Magaga dam	wheat, rice, vegetables maize
xii)	Zekarawa		40			Kafin Chiri dam	wheat, vegetables, maize
xiii)	Kusalla		30	30		Kusalla dam	wheat, vegetables, maize
xiv)	Wawan Rafi		20		1963	Pumping from natural lake Wawan Rafi	vegetables
xv)	Gwanda		10	10		borehole	orchard, vegetables
16.	Katsina						
i)	Makera		300	100		Pumping from Karaduwa river	wheat
ii)	Deberam		240	240		dam	wheat, vegetables, maize
iii)	Mashigi		150			dam	wheat, maize
iv)	Gagere		100	0		weir across Gagere river	
v)	Maruwa		100	76	1974	Pumping from Lake Maruwa	wheat, maize, vegetables
vi)	Mangwal		100			dam	wheat, maize
vii)	Gangagara		100	0		Gangara dam	wheat, vegetables
viii)	Raddewa		100	50		Raddawa dam	wheat, maize, vegetables
ix)	Koza		40	40		dam	wheat, vegetables

Table A. 3 cont'd.

State	Schemes	Owner if not State	Proposed area (ha)	Area irrigated by Dec. 1992 (or by Dec. 1980)	Year of 1st irrigation	Source of water	Crops
x)	Tsadoji		40	40		shallow groundwater	wheat, vegetables
xi)	Nasarawa		40	28	1972	Nasarawa dam across Daro river	wheat, tomatoes
xii)	Bakori		30	30		dam	wheat, maize, vegetables
xiii)	Maska		30			Maskadam	maize, vegetables
xiv)	Parvan Sanyi		16			dam	wheat
xv)	Karkio		10			dam	wheat, rice, vegetables
17.	Kebbi						
i)	Illo		120	0		dam across Illogour river	
ii)	Argungu Rice		20	20	1977	Pumping from Rima river	rice
iii)	Argungu Orchard		4	2	1973	water from Artesize borehole	orchard
iv)	Kebbi Orchard		4	2	1973	water from Artesize borehole	orchard
18.	Kogi						
i)	Aiyetoro Gbede		50	(3)	1974	dam across Ose stream	maize, vegetables
ii)	Serki Noma		50	(20)	1972	weir across Akpanioba stream	rice
19.	Kwara						
i)	Ajasse-Ipo		100	3(1987)	1976	Pumping from Osha river	citrus, vegetables
ii)	Oyan Nursery		30	2(1987)	1969	Pumping from Oyun river	citrus, vegetables
iii)	Shao		50	3(1987)	1972	Pumping from Imoru river	vegetables
iv)	Lafiagi		20	5(1987)	1963	Pumping from Ore river	vegetables
v)	New Yidi		20	2(1987)		weir across tributary of Asa river	citrus, vegetables
vi)	Oyun Offa Garden		3	5(1987)	1975	Pumping from Oyun river	vegetables
vii)	New Yidi		20	2(1989)	1976	weir across Ookun river	vegetables
viii)	Obbo-Ile		90	12(1992)	1975	Pumping from Iraji river	maize, citrus, vegetables

Table A. 3 cont'd.

State	Schemes	Owner if not State	Proposed area (ha)	Area irrigated by Dec. 1992 (or by Dec. 1980)	Year of 1st irrigation	Source of water	Crops
ix)	Moshe		50	5(1992)	1992	Pumping from Moshe river	vegetables, maize, sugar cane
x)	Passa Elepo		60	17(1992)	1989	dam across Passa river	maize vegetables
xi)	Bussamu		60	9(1992)	1989	dam across Bussamu river	maize vegetables, cowpea
xii)	Aderan		40	10(1992)	1992	Pumping from Awun river	maize, vegetables, sugar cane
xiii)	Liara-Isin		20	5(1992)	1992	Pumping from Oyin river	maize, vegetables
xiv)	Wurumar		10	6(1992)	1979	weir across Wuruma river	vegetables, sugar-cane, cotton
20.	Lagos						
i)	Itoikin	OORBRDA	315	141	1979	Pumping from Aye river	rice, maize, vegetables
ii)	Igbonla	OORBRDA	150	130			rice
21.	Niger						
i)	Dogonari	UNRBRDA	320	320		Pumping from Kanji lake	maize, vegetables
ii)	Nasko	UNRBRDA	200			dam across Shdogulbi river	maize, rice, vegetables
iii)	Loguma		125	(85)	1956	weir across Emi river	
iv)	Kuta		80	10	1981	Kuta dam across Rafin Kuta river	maize, vegetables
v)	Lioji		80	40		Pumping from Lioji river	rice, vegetables
vi)	Zara		80	0	1980	weir across Bagoma river	rice, wheat, maize, vegetables
vii)	Toroko		80		1954	weir across Kanko river	rice
viii)	Baratsu		73	(49)	1980	weir across Eji river	rice
ix)	Papiri		50	0		Pumping from Kanji lake	cowpea, vegetables
x)	Gusoro		40	0		Pumping from Kaduna river	wheat, maize, vegetables

Table A. 3 cont'd.

State	Schemes	Owner if not State	Proposed area (ha)	Area irrigated by Dec. 1992 (or by Dec. 1980)	Year of 1st irrigation	Source of water	Crops
xi)	Ma'Agba		40	20	1980	Pumping from Ewan river	rice
xii)	Chanchaga		10	6	1979	Pumping from Chanchaga river	vegetables
xiii)	Suleja		8	6	1972	Pumping from IKU river	vegetables
xiv)	Abuja		6	6		Pumping from Yiku river	vegetables
xv)	Tamania		6	1		Pumping from Olive river	vegetables
xvi)	Lanzu		6	6	1974	Pumping from Lanzu river	vegetables
22.	Ogun						
i)	Oke-Odan	OORBRDA	400	12		Pumping from Yewa river	maize, vegetables
ii)	Abeokuta	OORBRDA	20	28		Lekan Are dam across Ogun river	vegetables
iii)	Ago-Iwoye, Ibi-Ade, Ado-Odo, Ijebu-Ode, Ikenne-Remo, Ilewo-Ilaro, Odeda, and Sawonjo, all 10ha each planned for Horticultual crops and none was completed in 1992.						
23.	Ondo						
i)	Apariko	BORBRDA	250	0		dam across Apariko river	
ii)	Owena		200	0		dam across Owena river	
24.	Oshun						
i)	Okuku	OORBRDA	400	0		intake on Erinle river	
ii)	Oogi	OORBRDA	350	0		intake on Erintu river	
25.	Oyo						
i)	Eniosa	OORBRDA	250	10	1992	Pumping from small dam	maize, vegetables
ii)	Asa	OORBRDA	400	0		intake on Oye river	

Table A. 3 cont'd.

State	Schemes	Owner if not State	Proposed area (ha)	Area irrigated by Dec. 1992 (or by Dec. 1980)	Year of 1st irrigation	Source of water	Crops
26.	Plateau						
i)	Ganawuri		400	16	1969	Pumping from Gawasai river	wheat, rice, maize, vegetables
ii)	Gidan Adamu		100	20		Pumping from Yashi river	maize, vegetabies
iii)	Bokkos		100	0		dam	
iv)	Awuma		70	20		Pumping from Awuma river	
v)	Riyom		40	16	1965		rice, maize, vegetables, citrus
vi)	Ta Hoss		40	40	1962	natural spring near railway line	rice, maize, vegetables
vii)	Wase		40	24	1968	Pumping from Wase river	maize, vegetables
viii)	Vom garden		25	24	1972		maize, vegetables
27.	Rivers						
i)	Kpong	NDBRDA	200	90		Pumping from Anodoni river	rice
ii)	Otuokpoti	"	100	50		Pumping from Nun river	rice
iii)	Anyama-Ogbia	"	100	100		Pumping from Nun river	rice
28.	Sokoto						
i)	Goronyo		120	120	1968	Pumping from Rima river	wheat
ii)	Tungan Tudu		120	120	1967	natural spring fills reservoir all night, water used to irrigate during the day	wheat, rice
iii)	Bakura		100	100	1963	Pumping from Natu lake	rice, orchad
iv)	Sokoto Garden		2	0		borehole	vegetables
v)	Sokoto Orchad		2	2	1970	Pumping from Sokoto river	orchad
vi)	Sarki (25ha), Biniasa (4ha) abandoned before 1992						

Table A. 3 cont'd.

State Schemes	Owner if not State	Proposed area (ha)	Area irrigated by Dec. 1992 (or by Dec. 1980)	Year of 1st irrigation	Source of water	Crops
29. Taraba						
i) Donga	URBRDA	150	32			rice, maize
30. Yobe						
i) Mugura		120	28		Pumping from Yobe river	rice, wheat
ii) Bolaram					Pumping from Yobe river	wheat
iii) Wachakal					Pumping from Bunum Gana river	wheat, vegetables

Appendix A.4

Table A. 4: Large and Medium-scale Irrigation Schemes

State Schemes	Owner if not State	Proposed area (ha)	Area irrigated by Dec. 1992 (or by Dec. 1980)	Year of 1st irrigation	Source of water	Crops
1 Abia						
i) Igbere		1,300	250		Igbere dam across Igwu river	rice
2. Adamawa						
i) Savannah Sugar Estate	SSC	13,600	6,000	1978	Kiri dam across Gongola river	sugar cane
3. Akwa Ibom						
i) Mbiabet		3,000	300		pumping from Cross river	rice, vegetables, maize
ii) Ekoi, Ikono		500	30		pumping from Enyang river	
4. Anambra						
i) Lower Anambra, Omo	AIRBRDA	5,000	3,000	1987	pumping from Anambra river	rice
5. Bauchi						
i) Dadin Kowa	UBRBRDA	38,000	0	1979	Dadin Kowa dam across Gongola river	
ii) Bagel	UBRBRDA	5,700	10	1978	pumping from Bagel river	vegetables, maize, wheat, rice
iii) Balanga		2,800	2,200		Balanga dam across Balanga river	wheat maize, vegetables
iv) Kaltiingo		2,300			dam across Kaltango river	
v) Galala	HURBRDA	1,500	0		dam across Galala river	
vi) Waya	UBRBRDA	2,000	0		Waya dam across Waya river	
vii) Katini		1,200				
6. Benue						
i) Naka	LBRBRDA	600	50		dam across Ana river	
ii) Umogidi	LBRBRDA	500	0		dam across Ogbabakpa river	
iii) Koriko		1,000	15	1972	pumping from Oyimora river	maize, vegetables, sugar cane

Table A. 4 cont'd.

State	Schemes	Owner if not State	Proposed area (ha)	Area irrigated by Dec. 1992 (or by Dec. 1980)	Year of 1st irrigation	Source of water	Crops
iv)	Obagaji	"	1,000	0		pumping from Benue river	
v)	Makurdi	"	1,500	150		pumping from Benue river	rice, maize, vegetables
vi)	Ejule-Ojebe		1,500				
vii)	Oguna		1,000	8	1972	pumping from Benue river	maize, vegetables, sugar cane
viii)	Adi		1,000	24	1977	Pumping from Katsina-Ala river	rice, maize, vegetables
ix)	Katsina-Ala River Project	LBRBRDA	10,000				
	a) Dura			18	1972		
	b) Jato-Aka			100	1975		
	c) Nyitse			25	1980		
7.	Borno						
i)	South Chad	CBRDA	67,000	7,000	1979	pumping from Lake Chad	wheat, maize, vegetables
ii)	Ebeji		4,000	140(1963)	1963	pumping from Ebeji river	wheat
iii)	Jere Bowl Rice	CBRDA	1,300	0		dam across Ngadda river	
iv)	Gashua	"	1,000	250		pumping from Yobe river	
v)	Baga (Kirenowa) polder	"	20,000	500	1979	pumping from Lake Chad	
vi)	Jaffi		800	30	1987	pumping from Yobe river	
vii)	Gari Abdullahi		750			pumping from Gongola river	
viii)	Komadugu-Yobe						
	a) Yau		420	280(1965)	1960	pumping from Komadugu-Yobe river	wheat, rice
	b) Daya		1,200	13(1961)	1961	"	wheat, rice
	c) Abadam		1,200	344(1964)	1964	"	wheat, rice

Table A. 4 cont'd.

State Schemes	Owner if not State	Proposed area (ha)	Area irrigated by Dec. 1992 (or by Dec. 1980)	Year of 1st irrigation	Source of water	Crops
8. Cross River						
i) Adim		1,000	0		stream diversion	rice
ii) Ukum		1,000	0		dam across Aya river	rice
iii) Ofodun		800	0		pumping from stream	rice
iv) Idomi		600	0		stream diversion	rice
v) Akim Akim		500	0			rice, vegetables
vi) Assiga (800ha proposed) abandoned after the dam failed						rice, vegetables
9. Delta						
i) Ewu	NDBRDA	1,500	100		pumping from Forcados river	rice, vegetables
ii) Ewulu	BORBRDA	1,000	30		pumping from Asinru river	
10. Edo						
i) Ilushi	BORBRDA	5,000	100		pumping from Niger river	rice, vegetables
11. Enugu						
i) Uzo-uwani		1,000	315	1964	weir across Obina river	rice
ii) Adani		1,800	350		pumping from Anambra river	rice
12. Jigawa						
i) Agufa		700	0		pumping from Hadejia river	wheat, rice, vegetables
ii) Jahun		700	100		pumping from Hadejia river	wheat, rice, vegetables
iii) Abir		1,000	130		pumping from Hadejia river	wheat, rice, vegetables
iv) Jakarade		2,000	160		dam across Gari river	wheat, rice, vegetables

Table A. 4 cont'd.

State Schemes	Owner if not State	Proposed area (ha)	Area irrigated by Dec. 1992 (or by Dec. 1980)	Year of 1st irrigation	Source of water	Crops
v) Hadejia Valley	HRBRDA	12,500	500	1960	Challawa Gorge, Tiga	Wheat, rice, vegetables
13. Kaduna						
i) Galma		610	55	1970	pumping from Galma river	wheat, maize, rice, vegetables
ii) Tubo		620	100		pumping from Tubo river	wheat, vegetables
iii) Kangimi		1,900	120		dam across Kano river	maize, wheat, tomatoes
14. Kano						
i) Kano River Project Phase 1	HURBRDA	22,000	14,000	1971	Tiga dam across Kano river	wheat, maize, vegetables
ii) Kano River Project Phase 11	HURBRDA	4,000	0		Challawa Gorge dam across Challawa river	wheat, maize, vegetables
iii) Gari		4,200	950	1975	Gari dam across Tomas river	wheat, rice, vegetables
iv) Tomas		2,300	400	1976	Tomas dam across Gari river	wheat, rice, vegetables
v) Jakara		2,000	50	1975	Jakara dam across Jakara river	wheat, vegetables
vi) Watari		1,900	70		Watari dam across Watari river	wheat, vegetables
vii) Kafin Chiri		600	0		Kafin Chiri dam across Magaga river	
viii) Gwarzo		600	0		Magaga dam across Guzuguzu river	
ix) Guzuguzu		530	0		Guzuguzu dam across Guzuguzu river	

Table A. 4 cont'd.

State	Schemes	Owner if not State	Proposed area (ha)	Area irrigated by Dec. 1992 (or by Dec. 1980)	Year of 1st irrigation	Source of water	Crops
15.	Kastina						
i)	Zobe Dam	SRRBRDA	5,200	100		Zobe dam across Karaduwa river	
ii)	Jibiya dam	SRRBRDA	3,000	500		Jibiya dam across Gada river	
iii)	Ajiwa		1,900	500		Ajiwa dam	maize, vegetables
iv)	Kwande-Are		600	0			wheat, maize, vegetables
16.	Kebbi						
i)	Zauro Polder	SRRBRDA	11,200	50			
ii)	Wara		2,000	200		pumping from Kainji lake	rice, wheat
iii)	Gafara		500	50		"	rice, wheat
17.	Kogi						
i)	Oguma	LNBRDA	500	100		pumping from Benue river	
ii)	Adorun	'	500	80		intake on Oweh river	
iii)	Geriyan	'	500	20(1987)		pumping from Gurara river	maize, tobacco, vegetables
iv)	Otube		1,000	3	1977	pumping from Okwo river	maize, vegetables
v)	Oforachi	LNBRDA	1,500	100		pumping from Ofu river	
vi)	Ejule Ogebe	"	2,500	200		pumping from Ota lake	
vii)	Ejiba/Omi	"	4,100	6(1987)		pumping from tributary of Kampe river	maize, cowpea, vegetables
18.	Kwara						
i)	Oke-Oyi	LNBRDA	500	90		weir across Oshin river	maize, cowpea, vegetables
ii)	Oro-Ago	LNBRDA	500	80		pumping	
iii)	Ero	SWA	500	60		dam	
iv)	Omu-Aran	LNBRDA	1,300	230		dam across Areja river	vegetables

Table A. 4 cont'd.

State Schemes	Owner if not State	Proposed area (ha)	Area irrigated by Dec. 1992 (or by Dec. 1980)	Year of 1st irrigation	Source of water	Crops
v) Duku-Lade		2,000	50	1959	weir across Duku stream	rice
vi) Ikale	LNBRDA	2,700	300		intake	
vii) Tada/Shonga	LNBRDA	3,000	11(1987)		pumping from Niger river	rice, maize, cowpea, vegetables
viii) Bacita sugar estate	NCS	5,690	5,690	1963	pumping from Niger river	sugar cane
19. Niger						
i) Gafara		800	150		pumping from Kainji lake	
ii) Tungan Kawo	UNRBRDA	800	400	1998	Tungan Kawo dam across Ubandawaki river	
iii) Badeggi		830	830		pumping from Musa river	rice
iv) Guzan		900	400	1995	pumping from Yiko river	rice
v) Edo/Lapai		930	100	1970	pumping from Etswan river	rice
vi) Rabbah		1,000	110	1973	pumping from Niger river	rice
vii) Edozhigi		1,000	750	1970	intake on Kupanko Ejiko river	rice
viii) Bakogi		2,000	100	1951	pumping from Bakogi river	rice
ix) Swashi Valley	UNRBRDA	2,700	520	1976	dam across Swashi river	rice, maize, vegetables
x) Doko	"	3,000	350		pumping from Kaduna river	rice, maize, vegetables
xi) Kotangora	"	12,400	0		dam across Kotangora river	
20. Ogun						
i) Lower Ogun (Mokoloki)	OOBRDA	12,000	340		dam across Oyan river	
ii) Otta	OOBRDA	1,000	50		Pumping from Ore river	
iii) Eyinwa	"	1,000	100		intake on Ore river	
21. Oyo						
i) Igbojaiye	OOBRDA	2,000	10	1992	dam across Oye river	maize, vegetables

Table A. 4 cont'd.

State	Schemes	Owner if not State	Proposed area (ha)	Area irrigated by Dec. 1992 (or by Dec. 1980)	Year of 1st irrigation	Source of water	Crops
ii)	Ofiki		2,000	54	1992	dam across Ofiki river	maize, vegetables
iii)	Sepeteri	OOBRDA	2,000	24	1992	dam across Owunu river	maize, vegetables
iv)	Illora (100ha) and Ogbomosho (500ha) farm settlement schemes were being planned in 1992						
22.	Plateau						
i)	Awe	LBRBRDA	500	0		Pumping from Dep river	
ii)	Keffi Wambai		500	24	1969	Pumping	rice, maize, vegetables
iii)	Shendam	LBRBRDA	1,000	0		dam	sugar cane
iv)	Longkat	"	1,500	200	1973	Pumping from Shemankar river	rice, maize, vegetables
v)	Doma	"	2,000	100	1977	Pumping from Ohina river	maize, vegetables, banana, tobacco
vi)	Dep	"	2,200	150	1970	Pumping from Dep river	rice, maize, vegetables
vii)	Tde Dam/Mada River	"	30		1977	Pumping from Mada river	rice, maize, banana
23.	Rivers						
i)	Peremabiri	NDBRDA	2,500	350	1979	Pumping from Nun river	rice
24.	Sokoto						
i)	Kware		800	120	1926	Pumping from Rima river	rice, wheat, vegetables
ii)	Kalmalo		800	230	1973	Pumping from Kalmalo lake	rice, vegetables
iii)	Wumo		1,200	800	1973	dam across Rima river	wheat, rice onion

Table A. 4 cont'd.

State	Schemes	Owner if not State	Proposed area (ha)	Area irrigated by Dec. 1992 (or by Dec. 1980)	Year of 1st irrigation	Source of water	Crops
iv)	middle Rima valley	SRRBRDA	5,300	100		Goronyo dam	wheat, rice onion
v)	Bakolori	SRRBRDA	23,000	4,200	1978	Bakolori dam across Sokoto river	wheat, maize, cotton, vegetables
25.	Taraba						
i)	Taraba River Project	UNRBRDA	1,000	35		Pumping from Gongola river	
ii)	Gassol	UNRBRDA	3,000	200		Pumping from Gongola river	wheat
26.	Yobe						
i)	Kushimaga	CBRDA	500	200		Pumping from Gongola river	
ii)	Garin Abdullahi		750			Pumping from Gongola river	wheat
iii)	Jeffi		800			Pumping from Ndivana river	wheat
iv)	Ngalda	CBRDA	1,000	200		Pumping from Gongola river	
v)	Ngabu		2,000			Pumping from Hawal river	rice, wheat

Subject Index

Name Index

www.ingramcontent.com/pod-product-compliance
Ingram Content Group UK Ltd.
Pitfield, Milton Keynes, MK11 3LW, UK
UKHW020125290926
15471UKWH00003B/293

9 789780 294021